# Efficient Integration of 5G and Beyond Heterogeneous Networks

Zi-Yang Wu · Muhammad Ismail ·
Justin Kong · Erchin Serpedin ·
Jiao Wang

# Efficient Integration of 5G and Beyond Heterogeneous Networks

Zi-Yang Wu
Department of Electrical and Computer
Engineering
Northeastern University
Shenyang, Liaoning, China

Justin Kong
United States Army Research Laboratory
Adelphi, MD, USA

Jiao Wang
Department of Electrical and Computer
Engineering
Northeastern University
Shenyang, Liaoning, China

Muhammad Ismail
Department of Computer Science
Tennessee Technological University
Cookeville, TN, USA

Erchin Serpedin
Department of Electrical and Computer
Engineering
Texas A&M University
College Station, TX, USA

ISBN 978-981-15-8528-9        ISBN 978-981-15-6938-8   (eBook)
https://doi.org/10.1007/978-981-15-6938-8

This Springer imprint is published by the registered company Springer Nature Singapore Pte Ltd.
The registered company address is: 152 Beach Road, #21-01/04 Gateway East, Singapore 189721,
Singapore

# Preface

Integration of high-frequency air interfaces into 5G and beyond (5G+) wireless networks can release the congested Radio Frequency (RF) band and support the increasing demand for high data rates. This requires utilization of new communication bands such as millimeter wave (mmWave), terahertz, and visible light. The aforementioned high-frequency bands, however, present challenging wave propagation characteristics that call for special measures to ensure their efficient integration within the 5G+ heterogeneous networks (HetNets). This brief investigates planning and operational challenges of 5G+ HetNets and presents viable solutions for integration of one example of high-frequency band, namely, the visible light band. This focus on visible light is motivated by the mature development and pervasive deployment of illumination infrastructure.

From a planning perspective, the existing literature has not addressed yet joint planning strategies that specify the optimal Base Station (BS) densities for mixed RF and optical (visible light) networks. Such desirable joint planning strategies should satisfy the users' Quality-of-Service (QoS) in terms of coverage requirements and exhibit minimum HetNet power consumption. Specifying the optimal BS densities helps to set boundaries to discern when integration of both optical and RF networks is beneficial. In this brief, we illustrate how tools from stochastic geometry and optimization theory can be employed in order to define such cooperative boundaries.

From an operational viewpoint, an important communication scenario involves vertical handovers where mobile users can switch from the optical BS to the RF BS whenever the Line-of-Sight (LOS) link of the optical channel is absent. Such a scenario is very common in indoor mobile environments. A wise handover policy would not trigger lots of handovers for short absence of LOS links in the optical channels as this would increase the overall network latency. Data-driven approaches can be adopted to develop such a handover policy as deterministic optimization tools would stand inefficient due to abrupt link outages and stochastic optimization tools could be too conservative. To develop a data-driven handover policy, several challenges need to be addressed: (1) Unavailability of realistic data for mobile indoor optical wireless channels. This challenge calls for a realistic indoor human mobility model to overcome the practical limitations of existing models. Therefore, a realistic indoor mobility framework is developed by employing a semi-Markov renewal process to capture the return tendency and bounded Lévy walk models at a macroscopic scale as well as shortest paths, steering forces, and random orientations at a microscopic scale. Using this mobility model, a wireless channel dataset is built for mobile indoor environments. (2) Prediction of LOS link status in sparse optical channels. The optical channel dataset can be further exploited to develop a link predictor to forecast the absence of LOS links in optical channels and to decide whether a handover decision should be made or not. The challenge is that due to the sparsity nature of the optical LOS link, conventional sequence-to-sequence regression techniques would fail to carry out useful predictions of the link quality. As such, this brief presents novel methods for wireless channel abstraction and densification to enable the reuse of the conventional Long Short-Term Memory (LSTM)-based Recurrent Neural Networks (RNNs) for efficient prediction of optical LOS links. Given the channel dataset and the status prediction of the optical link, a Reinforcement Learning (RL) strategy is proposed to learn the environment dynamics and decide whether a vertical handover is needed to minimize the network latency.

Another operational scenario that is investigated herein brief involves multi-homing support where the resources offered by the optical and RF networks are aggregated in order to benefit of the reliability and high throughput of HetNets. The challenge faced is that the RF and optical channels vary at different timescales, and hence resource allocation should be carried out at two different timescales to reduce the overall computational complexity. This challenge is addressed by adopting a two-timescale RL strategy to specify the power allocations on the RF and optical channels to satisfy the users' target QoS.

Thanks to the holistic planning and operational strategies presented in this brief, the reader will receive in-depth knowledge on how to integrate future coexisting networks at high-frequency bands in a cooperative manner to enable reliable and high-speed 5G+ HetNets.

Shenyang, China                                                                    Zi-Yang Wu
Cookeville, USA                                                              Muhammad Ismail
Adelphi, USA                                                                          Justin Kong
College Station, USA                                                          Erchin Serpedin
Shenyang, China                                                                        Jiao Wang
May 2020

# Contents

# Abbreviations

| | |
|---|---|
| 3D | Three dimensional |
| 5G+ | The fifth generation and beyond |
| AC | Alternating current |
| ACO-OFDM | Asymmetrically clipped optical orthogonal frequency division multiplexing |
| AI | Artificial intelligence |
| APC | Area power consumption |
| AWGN | Additive white Gaussian noise |
| BS | Base station |
| CCDF | Complimentary cumulative distribution function |
| CDF | Cumulative distribution function |
| CIR | Channel impulse response |
| CNN | Convolutional neural network |
| DC | Direct current |
| DCO-OFDM | DC-biased optical orthogonal frequency division multiplexing |
| DL | Downlink |
| ECDF | Empirical cumulative distribution function |
| FCL | Fully connected layer |
| FOV | Field-of-view |
| HetNets | Heterogeneous networks |
| IR | Infrared light |
| LED | Light-emitting diode |
| Li-Fi | Light-Fidelity |
| LOS | Line-of-sight |
| LSTM | Long short-term memory |
| MBS | Macrocell base station |
| MIMO | Multiple-input multiple-output |
| mmWave | Millimeter wave |
| MSE | Mean-squared error |
| NLOS | Non-line-of-sight |

| | |
|---|---|
| NOMA | Non-orthogonal multiple access |
| ORWP | Orientation-based random waypoint |
| OWC | Optical wireless communication |
| PAR | Peak-to-average ratio |
| PD | Photodiode |
| PDF | Probability density function |
| PPP | Poisson point process |
| PSD | Power spectral density |
| QoS | Quality-of-service |
| RAT | Radio access technology |
| ReLU | Rectified linear unit layer |
| RF | Radio frequency |
| RL | Reinforcement learning |
| RNN | Recurrent neural network |
| RWP | Random waypoint |
| SBS | Small cell base station |
| SC-FDE | Single-carrier frequency-domain equalization |
| SINR | Signal-to-interference-plus-noise ratio |
| UE | User equipment |
| UL | Uplink |
| VL | Visible light |
| VLC | Visible light communication |
| WLAN | Wireless local area network |

# Chapter 1
# Introduction: Challenges in 5G+HetNet Integration

**Abstract** The current developments in the fifth generation and beyond (5G+) mobile networks focus on moving from the increasingly congested radio frequencies (RF) to higher frequency bands such as millimeter wave (mmWave), Terahertz, and Visible Light bands. Among these high frequency options, visible and infrared lights represent a strong candidate owing to the widely deployed lighting systems. Due to their distinct wave propagation characteristics, efficient integration of RF with such high frequency bands is needed for better reliability and high throughput. This introductory chapter presents an overview on RF and light (optical)-based wireless networks. This chapter highlights existing research efforts along with the limitations that should be addressed from planning and operational perspectives to enable a smooth integration of RF and optical networks in mobile environments.

## 1.1 Wireless Medium in 5G+ HetNets

Propagation of electromagnetic waves with wavelengths in the order of millimeter (mm) to nanometer (nm) has recently attracted considerable attention as it represents a promising solution to decongest the radio frequency (RF) spectrum and to implement the unprecedented services to be offered by the fifth generation and beyond (5G+) mobile networks. In this context, millimeter wave (mmWave), Terahertz waves, Infrared Light, Visible Light, and Ultraviolet Light have been all extensively investigated as potential communication options in the 5G+ mobile networks [1–5]. However, the vulnerability of high frequency waves to blockages, caused by the deterioration of diffraction abilities, hinders their adoption in practical scenarios that assume mobility. Fortunately, the integration of RF and higher frequency bands empowers a seamlessly heterogeneous coverage whereby intelligent handover and multi-homing strategies can improve the user's experience in mobile scenarios. Specifically, two kind of connections can be established in such heterogeneous networks (HetNets). The first is a multi-mode connection where a user connects to a single type of network (either an RF or a high frequency band network) at a time and implements vertical handovers between the two networks whenever necessary. The second is a multi-homing connection where a user connects to both networks

Z.-Y. Wu et al., *Efficient Integration of 5G and Beyond Heterogeneous Networks*,
https://doi.org/10.1007/978-981-15-6938-8_1

1

simultaneously and aggregates their resources to support data hungry applications. Such a HetNet is very attractive as it benefits from the large capacity available at the high frequency bands and the high reliability of RF networks.

Thanks to the mature development and pervasive deployment of solid state lighting, especially visible light-emitting-semiconductors, light-based networking represents a powerful technology that was adopted in numerous applications [6, 7]. This technology is referred to as visible light communication (VLC), and represents a sub-component of broader optical wireless communication (OWC) field. Because visible light wavelengths are distributed between 380 nm and 780 nm, VLC is very sensitive to blockages of line-of-sight (LOS) links. Similar to mmWave and Terahertz communications, a light base station (BS) provides limited coverage and thus it gives rise to a femto/atto-cell. Traditionally, the conventional RF medium has been frequently considered as an umbrella radio access technology (RAT) to offer macro BS coverage in HetNets [8]. However, when deploying such HetNets in mobile scenarios, many challenges need to be addressed from both planning and operational aspects. Therefore, we resort to a holistic solution that assumes RF-optical-based 5G+ Het-Nets. Such a framework can be adopted to other candidate high frequency bands. A useful observation is the fact that light represents the medium with the shortest wavelength among the aforementioned high frequency bands. Thus, if we can provide a viable integrative solution for VL with RF, then the other candidate bands with better diffraction capabilities than light can benefit from the same proposed solution.

The next two sections elaborate more on the planning and operational challenges faced by the integration of RF and optical networks. A short summary is then provided on the remaining chapters of this brief that is dedicated on presenting feasible solutions to current integration challenges.

## 1.2  Challenges in HetNet Planning

The existing literature has overlooked development of joint planning policies to optimize BS densities for RF and optical networks in 5G+ HetNets. Such joint planning should meet the quality-of-service (QoS) requirements of users and in the same time minimize the HetNet power consumption.

In order to carry out the joint planning of VLC and RF (sub-)networks, the coverage outage probability of femto/atto-cells should be determined, as the joint placement of such femto/atto-cells along with the macro RF cell should provide complementary coverage. For optical (VLC) networks, the outage probability depends on the field-of-view (FOV) of the receiver's photodiode (PD). However, the existing literature does not consider a more practical and general setting where the half FOV can assume any value such as less than or equal to $90°$ but rather it assumes the fixed value of $90°$ [9]. In such a practical general setting when the half FOV is less than $90°$, the user receives optical signals only from optical BSs within a certain confined radius. Therefore, the user may not be associated with any optical BS if the distance from the user's location to the nearest optical BS is greater than this confined radius. Furthermore, unlike the case when half FOV is exactly 90°, the interference from neighboring optical BSs is weaker when half FOV is less than 90°. In this general setting, the traditional approach presented in [9], which estimates the probability density function (PDF) of the interference as a sum of Gamma distributions, cannot be directly used to determine the coverage outage probability because all the moments of interference are unknown. Since the derived outage probability will be used in the optimization framework for joint efficient deployment of RF and VLC BSs, an approximate outage expression that introduces low computational complexity is highly desirable.

Since capacity and cost efficiency are crucial features of future networks [10], selection of optimal BS densities for RF and VLC networks should not only provide coverage support but also should minimize the HetNet overall energy consumption. On the other side, the existing works dedicated to planning of energy-efficient networks focus their attention only on RF standalone networks. In addition, energy efficiency of RF-optical (VLC) HetNets is only considered in literature from an operational viewpoint and not from a planning perspective [11]. Hence, further in-depth investigation is required to plan energy-efficient RF-optical HetNets via joint optimization of BS densities of RF and VLC networks.

## 1.3 Challenges in HetNet Operation

From an operational perspective, currently, the communication strategies proposed for the high frequency (optical) links of the HetNet consider the classical signaling schemes adopted in RF stand-alone networks. However, the sensitivity of dynamic optical channels to blockages renders the classical signaling approaches unreliable. Under abrupt dynamic channel conditions, the deterministic optimization tools are useless, while the stochastic optimization methods may be too conservative in developing efficient resource management policies. Therefore, new challenges questioning the applicability of traditional approaches for resource management in 5G+ HetNets have been raised up. These challenges include:

1. Is the traditional random mobility model, e.g., [12], that is widely used in literature sufficient to capture the essential features of the mobile optical channel? The same concern applies to HetNets that involve other high frequency bands that share similar wave propagation characteristics with the optical frequency band. The answer to these concerns is no since random mobility models are not reliable in such settings. An easy justification for this is the fact that the studies dealing with sensitivity of LOS links to blockages focused their modeling scope on details such as movement and orientation of the user equipment (UE). The more important issue of how to properly model the UE mobility in highly sensitive channels remains a critical problem to be addressed.

2. Are the fluctuations of optical channels including the LOS link outages and signal recoveries predictable? If so, are the conventional methods such as stochastic and

time series-based models capable of carrying out such a prediction in a way that provides useful information for efficient resource management? The answer to these questions lie actually in the channel dynamics as reflected by the human mobility model discussed in the previous point.

Even if one could provide an answer to these questions, we would still face a set of practical operational challenges that include:

1. Is it manageable to pursue a dynamic programming technique in a complex (mobile) wireless environment? Such dynamic programming techniques, if manageable, can offer seamless connectivity in 5G+ HetNets.
2. How to leverage the predictive analysis in HetNet resource management?
3. How to manage the vertical and horizontal handovers in a way that guarantee the users' target QoS in a highly dynamic environment?
4. How to solve the multi-homing resource allocation problem in a HetNet containing media that vary at different timescales (e.g, optical versus RF channels)?

To address these challenges, new methodologies must be introduced. Recently, the field of machine learning has witnessed a rapid development especially due to the advances in deep learning and reinforcement learning technologies. Deep learning is capable of modeling any function [13], which makes it possible to capture the complex relationship between the network state and the objectives of network prediction and resource management. Furthermore, reinforcement learning allows to formulate strategies to optimize an anticipated reward based on the environment dynamics [14]. Thus, RL helps to develop smart resource management and intelligent network optimization policies.

## 1.3.1  Intelligent Handovers

Light propagation triggers environment-related handovers to evade blockages caused by surroundings and users' bodies. It is essential for handover decisions to guarantee QoS. However, sometimes handovers present a negative impact. This is observed in indoor settings when comparing the duration of a handover process with the duration of a single movement period. For example, a user's movement from one desk to another desk usually lasts a few seconds, whereas merely a single handover process could take hundreds of milliseconds. This implies that more than $\sim$10% of the movement duration would suffer from the handover latency. On the other hand, a long walk would induce several handovers among the light (optical) BSs along the way. Given the random orientations of UE, handovers are not deterministic and many of them are superfluous caused by ping-pong effects. Hence, a handover skipping policy should be designed based on the impact of user mobility on channel states.

Unfortunately, we have very limited access to the mobility information in indoor settings due to inaccurate measurements. Thus, mathematical mobility models were proposed, e.g., random waypoint (RWP) mobility model [15]. In [15], it is shown that the handover rate can be significantly reduced following a velocity-aware method that exploits the received optical power. However, focusing only on handover rate reduction is not sufficient to guarantee QoS, as network latency is affected not only by the handover rate but also by the inadequate data rates. Hence, in an RF-optical HetNet, the link assignment policy should consider the possibility when the handover could be triggered from a seemingly stable RF BS to a light BS with higher instantaneous outage risks but offering higher capacity. However, if the link assignment strategy aims to maximize only the achievable data rate, the outcome could be radical. In this case, a link assignment strategy may assign an UE to a BS with instantaneous high data rate but long outage duration rather than assigning the UE to another BS with lower instantaneous data rate but shorter outage duration.

Under realistic human mobility patterns, the UE orientation as well as its trajectory exhibit high dynamics and uncertainty. In addition, they are restricted to follow the statistical patterns induced by the indoor environment layout and human behavior. Due to these considerations, the feasibility of adopting in RF-optical HetNets conventional model-driven approaches for handover and resource management is highly questioned. On the other hand, employing closed-loop optimization strategies based on data-driven and reinforcement learning approaches is shown to represent a powerful tool to address the highly dynamic nature of the problem [16]. Hence, the overall decision process regarding link assignment can be designed based on predictive network analysis. In this context, predicting the channel gain (and hence potential capacity) or merely the channel events (i.e., likelihood of link outage and recovery) can facilitate the handover decision management to achieve low network latency and to satisfy a target QoS.

## *1.3.2 Channel Event Prediction*

Predicting the gain or status of an optical (or any high frequency band) wireless channel is quite different from its RF counterpart due the following reasons.

*Indoor Information Restrictions:* Given the inaccuracies in measurement of indoor locations along with the users' privacy concerns, mobility prediction models [15] are not useful to assist the handover management policy for indoor settings. Hence, the only information that can be used to evaluate the quality and sustainability of optical links is the channel gain data.

*Event Sparseness:* The LOS attenuation along with the outages due to UE random movements and burst blockages increase the channel dynamics and break the continuity of the channel gain sequence. As will be elaborated in this brief, the LOS outage probability of optical (high frequency) links exhibits an obvious spatio-temporal dependency on users' mobility pattern and environment layout. Frequent and long-term LOS outages associated with users' mobility cause sparseness in the channel gain, as there would be lots of abrupt "zeros" and many continuous "zeros" in the LOS channel gain. Therefore, conventional time sequence prediction algorithms will be paralyzed by such sparseness and high dynamic range.

*Prediction Overhead:* Since the channel gain/event prediction is conducted online to assist the handover decision management, it should present reduced computational complexity in order to be conducted fast. However, the prediction outcome is useful only if it presents a far ahead vision on the channel status. This should be at least far enough to cover both the prediction and handover delays. As such, the prediction capability should yield a foresight on the order of hundreds of milliseconds, which means several frames[1] ahead. Balancing the accuracy of such prediction with the reduced computational time requirement is a challenging task.

### 1.3.3   Data-Driven Optimization

The performance of data-driven approaches is mainly dominated by the quality of available datasets. Existing literature, however, lacks datasets for mobile optical wireless channels due to the extreme complexity and inaccuracy in data collection. This impedes the progress of data-driven approaches. Hence, it is imperative to develop a data generation mechanism to produce high quality relevant datasets.

The accuracy of human mobility representation is vital. If the user mobility is totally random without any constraints, it is impossible to predict the channel status. Simple random walk models, such as orientation-based random waypoint (ORWP) models, are widely-used in literature; although they fail to accurately characterize indoor mobility from the perspective of UEs. In an ORWP model, the moments where users are steering or changing their speed hardly depend on space (room layout) or time. Hence, the ORWP model fails to capture realistically the spatio-temporal progress of link outages.

Recent statistics reveal that on a macro spatio-temporal scale human mobility complies with return tendency and a bounded Lévy-walk model [17]; while on a micro-scale, it adheres also to a specific path selection and steering behavior [18]. In addition, random orientations of UE are also present at a micro-scale level [19]. These aspects collectively evolve into a multi-scale UE-centric mobility model, which unfolds a realistic spatio-temporal behavior in mobile optical communications. As will be explained later in this brief, there exist multiple peaks in the channel gain distribution, evolving over time as the UEs' move. Such a realistic mobility model empowers data-driven resource management strategies for mobile optical communication links and for their integration into 5G+.

### 1.3.4   Intelligent Multi-homing Support

Besides multi-mode connections, multi-homing support should also be investigated in 5G+ HetNets. In a multi-homing setting, resource management strategies are fur-

---

[1] A frame consists of a set of time slots.

ther challenged as in such HetNets the resource allocation decisions in different networks should operate at different timescales [20]. Specifically, in RF-optical Het-Nets, RF channels change more rapidly than optical (VLC) channels. Therefore, the power allocations in the RF and VLC networks should be conducted at different timescales. Unlike the multi-mode users where the achieved rate at a user is either the rate yielded by the RF network or by the VLC network [21], the achieved rate at a multi-homing user is the sum of the rates offered together by the RF and VLC networks. Hence, it is important to carefully determine the RF transmit powers in every time slot by considering the dynamics of the rates achieved by the VLC network. On the other hand, the power allocation at the VLC network should be carried out at the beginning of each frame by taking the users' mobility into account. Hence, two-timescale resource allocation policies should be developed to support multi-homing users in 5G+ HetNets.

## 1.4 The Road Ahead

Motivated by these challenges, we introduce in this brief a holistic solution that spans both the planning and operational aspects for efficient integration of RF and optical networks in 5G+ HetNets. A brief summary of the remaining chapters is given below.

**Chapter** 2: **Efficient Joint Planning of 5G+ HetNets**: This chapter focuses on the development of energy efficient HetNets that employ both RF and VLC BSs. More specifically, since the QoS and energy cost are key parameters in designing energy efficient networks, this chapter optimizes the RF and VLC BS densities to minimize the area power consumption (APC) by enforcing an outage probability constraint. Using stochastic geometry tools, approximations of the outage probability of VLC networks are first introduced. Leveraging the derived analytical results, a set of algorithms are presented to identify the optimal densities of RF and VLC BSs for energy efficient RF-optical HetNets. The proposed algorithms are computationally efficient since they reduce to employing one-dimensional search methods.

**Chapter** 3: **Realization and Dataset Generation of Mobile Indoor Channels**: This chapter focuses on the development of high-quality datasets for mobile optical wireless channels. These datasets are necessary for developing robust data-driven resource management strategies for RF-optical HetNets. An analytical framework to model the propagation channel characteristics in a mobile OWC network whose downlink relies on VLC and uplink is implemented via infrared light is presented. The proposed analytical framework generates trajectories that reflect the nature of human behavior and integrates both macro and micro mobility patterns. These patterns are then used within the proposed framework to model the spatio-temporal characteristics of optical wireless channels under long-term mobility.

**Chapter** 4: **Data-driven Handover Algorithm in Mobile 5G+ HetNets**: This chapter investigates vertical handovers in RF-optical HetNets. To ensure a reliable link with QoS guarantee in terms of network delay and throughput, vertical handovers are triggered within the HetNet. A data driven approach is adopted to reach optimal handover decisions based on the created dataset in the previous chapter. The proposed data-driven algorithm predicts abrupt outages in LoS optical links and evaluates the optical channel quality using a deep long-short-term-memory (LSTM)-based recurrent neural network (RNN). Given the resulting LOS link outage prediction in OWC, a reinforcement learning-based approach is presented to implement optimal vertical handover decisions with QoS guarantee. The presented handover decision algorithm learns to make a trade-off between the outage risk and the cost of excessive handovers, and achieves considerable mitigation on overall latency and handover rate under indoor mobility.

**Chapter** 5: **Data-driven Multi-homing Resource Allocation in Mobile 5G+ HetNets**: This chapter investigates mobile multi-homing HetNets composed of an RF BS and multiple VLC BSs, where users can aggregate resources from both RF and VLC BSs in the downlink. In RF-optical HetNets, RF channel gains vary faster than those of VLC channels due to the small scale fading. By leveraging multi-agent Q-learning to interact with the dynamics of wireless environments, we present an online two-timescale power allocation strategy that optimizes the allocated powers to ensure QoS satisfaction for RF-optical HetNets.

**Chapter** 6: **Conclusions**: This chapter summarizes the main findings in this brief and discusses future research directions.

All abbreviations used in this brief are listed in "List of Abbreviations".

# References

1. T.S. Rappaport et al., Millimeter wave mobile communications for 5G cellular: it will work! IEEE Access **1**, 335–349 (2013)
2. H. Song, T. Nagatsuma, Present and future of terahertz communications. IEEE Trans. Terahertz Sci. Technol. **1**(1), 256–263 (2011)
3. J.M. Kahn, J.R. Barry, Wireless infrared communications. Proc. IEEE **85**(2), 265–298 (1997)
4. H. Haas et al., What is LiFi? J. Lightw. Technol. **34**, 1534–1544 (2016)
5. Z. Xu, B.M. Sadler, Ultraviolet communications: potential and state- of-the-art. IEEE Commun. Mag. **46**(5), 67–73 (2008)
6. D. Tsonev et al., A 3-Gb/s single-LED OFDM-based wireless VLC link using a gallium nitride 'LED'. IEEE Photon. Technol. Lett. **26**(7), 637–640 (2014)
7. Z-Y. Wu et al., A linear current driver for efficient illuminations and visible light communications. J. Lightw. Technol. **36**(18), 3959–3969 (2018)
8. R. Zhang et al., Visible light communications in heterogeneous networks: paving the way for user-centric design. IEEE Wirel. Commun. **22**(2), 8–16 (2015). ISSN: 1536-1284
9. D.A. Basnayaka, H. Haas, Design and analysis of a hybrid radio frequency and visible light communication system. IEEE Trans. Commun. **65**, 4334–4347 (2017)
10. J.G. Andrews et al., What will 5G be? IEEE J. Sel. Areas Commun. **32**, 1065–1082 (2014)
11. M. Ismail et al., *Green Heterogeneous Wireless Networks*, 1st edn. (Wiley, Hoboken, 2016)
12. C.-L. Tsao et al., Link duration of the random way point model in mobile ad hoc networks, in *Proceedings of IEEE Wireless Communications and Networking Conference* (WCNC)
13. Y.L. Cun, Y. Bengio, G. Hinton, Deep learning. Nature **521**(7553), 436–444 (2015)
14. R.S. Sutton, A.G. Barto, *Reinforcement learning: An introduction* (MIT Press, Boca Raton, 2018)
15. X. Wu, H. Haas, Handover skipping for LiFi. IEEE Access **7**, 38369–38378 (2019)

16. Z. Wu et al., Data-driven link assignment with QoS guarantee in mobile RF-optical HetNet-of-Things. IEEE Internet Things J. 1–1 (2020)
17. M.C. Gonzalez, C.A. Hidalgo, A-L. Barabasi, Understanding individual human mobility patterns. Nature **453**(7196), 779 (2008)
18. C.W. Reynolds, Steering behaviors for autonomous characters, in *Game Developers Conference*, vol. 1999. Citeseer (1999), pp. 763–782
19. M.D. Soltani et al., Modeling the random orientation of mobile devices: measurement, analysis and LiFi use case. IEEE Trans. Commun. **67**(3), 2157–2172 (2018)
20. H.S. Chang et al., Multitime scale Markov decision processes. IEEE Trans. Autom. Control **48**, 976–987 (2003)
21. W. Wu, F. Zhou, Q. Yang, Adaptive network resource optimization for heterogeneous VLC/RF wireless networks. IEEE Trans. Commun. **66**, 5568–5581 (2018)

# Chapter 2
# Efficient Joint Planning of 5G+ HetNets

**Abstract** This chapter focuses on the deployment of energy efficient 5G and beyond (5G+) heterogeneous networks (HetNets) that employ both radio frequency (RF) and high frequency bands. Since the quality-of-service and energy cost are key parameters in designing energy efficient networks, in this chapter the base station (BS) densities for different frequency bands are optimized to minimize the area power consumption (APC) by taking into account an outage probability constraint. The considered setting assumes integration of RF BSs with optical BSs. Using stochastic geometry tools, approximations of the outage probability of optical visible light communication (VLC) networks, which are applicable to an arbitrary field-of-view (FOV) at photodiodes (PDs) and present low computational complexity, are first proposed. Leveraging the derived analytical results, a low complexity algorithm to find the optical BS density that minimizes the APC of VLC networks is then recommended. Reduced complexity algorithms that rely on one-dimensional searching approaches and that identify the optimal densities of RF BSs and optical BSs to secure energy efficiency of RF-optical HetNets are also proposed. Numerical simulations corroborate the tightness of the approximations proposed for the outage probability and confirm that the introduced algorithms exhibit almost identical performance as the algorithms that exhaustively search the optimal densities of BSs. It is shown that the RF-optical HetNets achieve a lower outage probability and a more reduced APC compared to the RF-only networks and VLC-only networks.

## 2.1 Introduction

Visible light communication (VLC) has recently attracted considerable attention as a promising solution to overcome the scarcity of radio frequency (RF) spectrum and to meet the high demands of wireless communication services [1–3]. However, since each VLC transmitter illuminates only a small confined area, standalone VLC networks may not support a user residing in a coverage hole. On the other hand, RF networks provide ubiquitous coverage with moderate data rates. In addition, VLC signals and RF signals do not interfere with each other. Motivated by these facts, RF-optical heterogeneous networks (HetNets), which can offer high data rates ubiquitously, have been widely explored lately [4, 5]. By exploiting stochastic geometry tools [6], this chapter focuses on the deployment of energy efficient RF-optical HetNets that minimize the area power consumption (APC) while guaranteeing a quality-of-service (QoS) requirement [7].

### 2.1.1 Background

During the past decade, many researchers have investigated VLC networks [8–13]. The authors in [8] derived upper and lower bounds in closed-form expressions for the channel capacity of VLC networks. Another closed-form upper bound for the channel capacity, which is tighter than the bound in [8], was introduced in [9]. The work in [10] introduced three adaptive modulation schemes for high speed VLC networks, i.e., DC-biased optical orthogonal frequency division multiplexing (DCO-OFDM), asymmetrically clipped optical OFDM (ACO-OFDM), and single-carrier frequency-domain equalization (SC-FDE). Also, high data rate achieving multiple-input multiple-output (MIMO) techniques for VLC networks were provided in [11]. The study in [12] examined zero-forcing precoding designs for multi-user MIMO VLC networks, and the authors in [13] assessed the bit-error-rate performance of VLC networks when non-orthogonal multiple access (NOMA) is adopted.

Recently, RF-optical HetNets have drawn significant attention due to the complimentary nature of RF networks and VLC networks. In [14], an intelligent algorithm to carry out a handover decision in RF-optical HetNets was presented. The works in [15, 16] developed dynamic load balancing schemes by taking user fairness into account. A two-stage base station (BS) selection technique exhibiting a near optimal throughput with a reduced complexity was introduced for RF-optical HetNets in [17]. The problem of power and bandwidth allocation for energy efficient RF-optical HetNets consisting of a single RF BS and a single VLC BS was considered in [18]. In addition, the authors in [19] investigated energy efficient subchannel and power allocation methods for software-defined RF-optical HetNets with fixed numbers of RF and VLC BSs.

By modeling the spatial distributions of BSs as Poisson point processes (PPPs), the performances of VLC networks and RF-optical HetNets have been analyzed in [20–23]. In [20], the signal-to-interference-plus-noise ratio (SINR) outage probability was assessed when the positions of optical BSs are driven by a PPP. Under the assumption that the locations of optical BSs and RF BSs follow two independent PPPs, the rate outage probability of RF-optical HetNets was characterized, and the minimum spectrum and power for RF networks to achieve certain rate requirements were examined in [21]. Also, the authors in [22] provided an analytical expression of the coverage probability of multi-user VLC networks by modeling the spatial distribution of active optical BSs as a thinned PPP. The studies in [20–22] assumed full FOV of 180° at the receiver's PD, i.e., half of FOV is equal to 90°, which is not practical [24, 25]. An approximation of the coverage probability of VLC networks with an arbitrary FOV was introduced in [23]. However, since the approximation in [23] resorts to Gil-Pelaez inversion theorem [26], the approximation requires a three-fold integration which presents a high computational complexity. As highlighted before, simple expressions to approximate the performance of VLC networks have not been reported in the existing literature.

The coverage probabilities of single-tier and multi-tier RF networks were derived in [27–29] assuming that the locations of BSs follow independent PPPs. The work in [30] showed that two-tier RF networks with BS sleeping can significantly reduce the network energy cost compared to single-tier RF networks. Optimizing the BS densities in multi-tier RF networks in the form of power consumption minimization and energy efficiency was investigated in [31]. The optimal BS densities that maximize the spatial throughput of multi-tier RF networks was explored in [32] by considering the traffic loads of individual tiers. For single-tier and two-tier RF networks, the authors in [33] derived the optimal BS densities and transmit powers that minimize the APC under coverage performance constraints. The study in [34] generalized the results of [33] by considering biased $K$-tier RF networks and three types of user association schemes.

The proliferation of light-emitting-diodes (LEDs) in public area illumination infrastructures has increased the interest for deployment of outdoor VLC networks [35–39]. In outdoor settings, optical BSs could be traffic lights, street lights, park lights, LED boards at railway stations, etc. [37]. The practical feasibility of outdoor VLC networks was investigated in [37–39] by considering the influence of sunlight. Another factor contributing to the deployment of VLC networks is the fact that VLC networks can be deployed with low cost as there is no need to construct new infrastructures. Since energy and cost efficiencies are critical factors in 5G and beyond (5G+) networks [40], VLC networks represent promising components for the future 5G+ networks.

In summary, the existing research suffers from the following limitations: (1) Previous works dealing with the deployment of energy efficient networks have focused only on RF networks. (2) Existing studies on energy efficient VLC networks focus only on operational aspects and completely overlook designing optimal planning strategies. (3) Joint planning of energy efficient RF-VLC HetNets should guarantee coverage requirements for all users. Existing studies on VLC outage probability either assume the special setting of 180° for FOV of PDs or rely on complex expressions for the general setting when FOV is less than or equal to 180°. To provide optimal joint planning strategies with low computational complexity, an approximate expression for VLC coverage outage probability is required. Therefore, this chapter focuses on designing energy efficient VLC networks and RF-VLC HetNets by optimizing the densities of optical BSs and RF BSs in outdoor environments.

### 2.1.2   Chapter Organization

The rest of this chapter is organized as follows: Section 2.2 introduces the network model of RF-optical HetNets and formulates the energy efficient APC minimization problem. Tight approximations of the outage probability of VLC networks are presented in Sect. 2.3. Section 2.4 introduces algorithms to identify the BS densities for energy efficient standalone VLC networks and RF-optical HetNets. Numerical simulation results that corroborate the efficiency of the introduced algorithms are presented in Sect. 2.5.

## 2.2   Network Model and Problem Formulation

Throughout this chapter, the following notations are used. $\mathbb{P}(A)$ denotes the probability of an event $A$ and $\mathbb{E}[X]$ represents the expectation of random variable $X$. The operators $\| \cdot \|$ and $\backslash$ indicate the Euclidean 2-norm and set difference, respectively. In addition, the bold notation is adopted to denote a point $\mathbf{x} \in \mathbb{R}^2$.

We investigate RF-optical HetNets where a user can communicate with either RF BSs or optical BSs. The horizontal locations of the optical BSs are modeled as a 2-D PPP $\Phi_V$ with density $\lambda_V$ and the height of the optical BSs is $L$. Two tiers are assumed for the RF network with macro cell BSs (MBS) and small cell BSs (SBS). The spatial distributions of the MBSs and SBSs are assumed to follow a 2-D PPP $\Phi_M$ with density $\lambda_M$ and a 2-D PPP $\Phi_S$ with density $\lambda_S$, respectively. PPPs $\Phi_V$, $\Phi_M$, and $\Phi_S$ are assumed to be mutually independent. The objective is to specify the optimal BS densities, $\lambda_V$, $\lambda_M$, and $\lambda_S$, that minimize the HetNet APC while satisfying the coverage outage requirements.

### 2.2.1   Area Power Consumption

We first focus the attention on the APC of RF-optical HetNets. The power consumption at an RF BS in $\Phi_i$ is expressed as $P_i^C + \Delta_i P_i$ for $i \in \{M, S\}$ where $P_i^C$, $\Delta_i$ and $P_i$ respectively stand for the fixed circuit power consumption, the slope of load-dependent power consumption and the transmit power at an RF BS in $\Phi_i$ [33, 34]. As the transmission power at an optical BS is the same optical power used for illumination, the fixed power consumption $P_V$ is only considered as the communication cost for optical BSs [18].[1] Then, the APC (in Watt/m$^2$) of the RF-optical HetNet is expressed as

---

[1] The fixed power consumption $P_V$ is for the case where optical BSs send modulated optical signals at full duty cycle. As the fixed power overestimates load-dependent communication powers, the approach in this chapter can be interpreted as the minimization of an upper bound of the APC with the load-dependent VLC power consumption.

$$P(\lambda_V, \lambda_M, \lambda_S) = \lambda_V P_V + \lambda_M \bar{P}_M + \lambda_S \bar{P}_S, \tag{2.1}$$

where $\bar{P}_M \triangleq P_M^C + \Delta_M P_M$ and $\bar{P}_S \triangleq P_S^C + \Delta_S P_S$. The objective of this chapter is to design energy efficient RF-optical HetNets that minimize the APC expression specified in (2.1) via optimal selection of $\lambda_V$, $\lambda_M$, and $\lambda_S$. This further requires the characterization of the outage probability of RF networks and VLC networks with limited FOV, a topic which is discussed next.

## 2.2.2 Preliminaries on Outage Probability of RF Networks

In this subsection, we review features of the outage probability of RF networks, which will be utilized in Sect. 2.4. From the stationarity of PPP, a typical user is assumed to be positioned at the origin $o$ and is associated with the RF BS that provides the highest average received power [29]. When the user is connected to a BS in $\Phi_i$ for $i \in \{M, S\}$, we can represent the resulting SINR as

$$\gamma_i = \frac{P_i h_{\mathbf{x}_R} \|\mathbf{x}_R\|^{-\alpha}}{\sum_{j \in \{M,S\}} \sum_{\mathbf{x} \in \Phi_j \setminus \mathbf{x}_R} P_j h_{\mathbf{x}} \|\mathbf{x}\|^{-\alpha} + \sigma_R^2} \quad \text{for } i \in \{M, S\},$$

where $\alpha$ denotes the path loss exponent, $\mathbf{x}_R$ accounts for the location of the associated BS and $h_{\mathbf{x}}$ represents the power gain of small scale fading channel between the user and the BS at $\mathbf{x}$, which is modeled as an exponential random variable with unit mean. The power of additive white Gaussian noise (AWGN) is represented through variable $\sigma_R^2$.

The outage probability is defined as the probability that the SINR is less than a certain threshold $\gamma_{\text{th}}$. The RF network outage probability $\mathcal{P}_o^R$ is expressed as

$$\mathcal{P}_o^R = \sum_{i \in \{M,S\}} \mathcal{A}_i \mathbb{P}(\gamma_i < \gamma_{\text{th}}), \tag{2.2}$$

where $\mathcal{A}_M$ ($\mathcal{A}_S$) denotes the probability that the typical user is connected to an MBS (SBS) and is expressed as [29]

$$\mathcal{A}_i = \frac{\lambda_i P_i^{2/\alpha}}{\lambda_M P_M^{2/\alpha} + \lambda_S P_S^{2/\alpha}} \quad \text{for } i \in \{M, S\}.$$

In addition, from [29], $\mathbb{P}(\gamma_i < \gamma_{\text{th}})$ is given by

$$\mathbb{P}\left(\gamma_i < \gamma_{\text{th}}\right) = 1 - \frac{2\pi\lambda_i}{\mathcal{A}_i} \int_0^\infty x \exp\left(-\frac{\gamma_{\text{th}}\sigma_R^2 x^\alpha}{P_i}\right) \exp\left(-\pi x^2 \sum_{j\in\{M,S\}} \lambda_j \left(\frac{P_j}{P_i}\right)^{2/\alpha}\right)$$

$$\times \mathbb{E}\left[\exp\left(-\frac{\gamma_{\text{th}} x^\alpha}{P_i} \sum_{\mathbf{x}\in\Phi_j\backslash\mathbf{x}_R} P_j h_{\mathbf{x}} \|\mathbf{x}\|^{-\alpha}\right)\right] \mathrm{d}x.$$

Using the probability generating functional property of PPP [6], we infer further that

$$\mathbb{E}\left[\exp\left(-\frac{\gamma_{\text{th}} x^\alpha}{P_i} \sum_{\mathbf{x}\in\Phi_j\backslash\mathbf{x}_R} P_j h_{\mathbf{x}} \|\mathbf{x}\|^{-\alpha}\right)\right] = \exp\left(-\pi x^2 \rho(\gamma_{\text{th}}, \alpha) \sum_{j\in\{M,S\}} \lambda_j \left(\frac{P_j}{P_i}\right)^{2/\alpha}\right),$$

where $\rho(\gamma_{\text{th}}, \alpha) \triangleq \gamma_{\text{th}}^{2/\alpha} \int_{\gamma_{\text{th}}^{-2/\alpha}}^\infty \frac{1}{1+u^{\alpha/2}} \mathrm{d}u$. Adopting the change of variable $t = x^2 P_i^{-2/\alpha}$, we obtain

$$\mathbb{P}\left(\gamma_i < \gamma_{\text{th}}\right) = 1 - \frac{\pi\lambda_i P_i^{2/\alpha}}{\mathcal{A}_i} \int_0^\infty \exp\left(-\gamma_{\text{th}}\sigma_R^2 t^{\alpha/2}\right)$$

$$\times \exp\left(-\pi t(1 + \rho(\gamma_{\text{th}}, \alpha)) \sum_{j\in\{M,S\}} \lambda_j P_j^{2/\alpha}\right) \mathrm{d}t.$$

Hence, the RF network outage probability $\mathcal{P}_o^R$ in (2.2) becomes

$$\mathcal{P}_o^R(\xi) = 1 - \pi\xi \int_0^\infty \exp\left(-\gamma_{\text{th}}\sigma_R^2 t^{\alpha/2}\right) \exp\left(-\pi t\xi(1 + \rho(\gamma_{\text{th}}, \alpha))\right) \mathrm{d}t, \qquad (2.3)$$

where $\xi \triangleq \lambda_M P_M^{2/\alpha} + \lambda_S P_S^{2/\alpha}$. It is shown that $\mathcal{P}_o^R(\xi)$ is a non-increasing function of $\xi$ [34, Lemma 3] and it converges to a certain value when $\xi$ goes to infinity [29, Corollary 2], i.e.,

$$\mathcal{P}_o^R(\xi) \xrightarrow[\xi\to\infty]{} \frac{\rho(\gamma_{\text{th}}, \alpha)}{1 + \rho(\gamma_{\text{th}}, \alpha)}. \qquad (2.4)$$

Figure 2.1 illustrates the outage probability of RF networks $\mathcal{P}_o^R(\xi)$ when $\sigma_R^2 = -100$ dBm and $\alpha = 4$. First, we can observe that $\mathcal{P}_o^R(\xi)$ is a non-increasing function of $\xi$, and it decreases as $\gamma_{\text{th}}$ decays. Since $\mathcal{P}_o^R(\xi)$ converges as $\xi \to \infty$, when a target outage probability is smaller than the converged value, the RF networks cannot fulfill the requirement on the outage probability. Also, when $\xi$ is large enough, increasing $\xi$ has only a negligible impact on $\mathcal{P}_o^R(\xi)$ while consuming additional energy. These facts motivate the search for energy efficient RF-optical HetNet that can achieve the target outage probability with optimized network energy consumption.

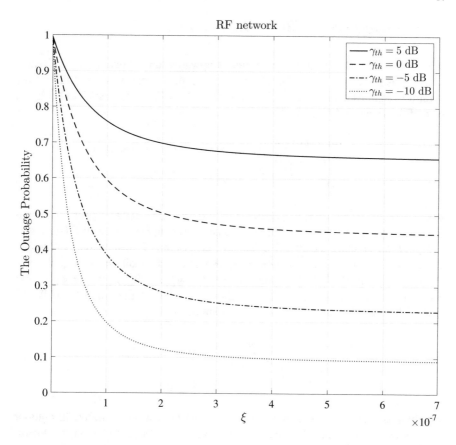

**Fig. 2.1**  Outage probability of RF networks $\mathcal{P}_o^R(\xi)$ as a function of $\xi$ [7]

In practical networks, a correlation between the locations of MBSs and SBSs may exist. For example, stochastic geometry approaches can be used to model the correlations present in the setting where SBSs are deployed farther than a minimum distance from MBSs, i.e., outside the exclusion zones of a given radius [41, 42]. Although some approximations on the performance were introduced in [41, 42], these approximations present complicated forms, and hence, they cannot be used to optimize the densities of MBSs and SBSs. Note that since the interference from SBSs is reduced with correlation, the outage probability in (2.3) is an upper bound for the outage probability of the networks with correlation. From this regard, the study presented in this chapter can be considered as a worst case scenario optimization.

### 2.2.3  Outage Probability of VLC Networks

It is assumed that the PD is facing vertically upwards[2] and optical BSs are directed downwards since the optical BSs could be street lights, park lights, and LED boards at railway stations that are generally tilted downwards. Then, for each VLC link, the gain of the channel is given by [21–23]

$$h = \frac{(m+1)A_{\mathrm{pd}}\eta}{2\pi(x^2 + L^2)} \cos^m(\theta_{tx}) G_f(\theta_{rx}) G_c(\theta_{rx}) \cos(\theta_{rx})$$
$$= \kappa(x^2 + L^2)^{-\frac{m+3}{2}}, \tag{2.5}$$

where $\kappa \triangleq (m+1)A_{\mathrm{pd}}\eta G_f(\theta_{rx}) G_c(\theta_{rx}) L^{m+1}/(2\pi)$, and $x$ and $L$ respectively denote the horizontal and vertical separations between a user and an optical BS. Here, $A_{\mathrm{pd}}$, $\eta$, $\theta_{tx}$ and $\theta_{rx}$ stand for the effective detection area of the PD, responsivity of the PD, angle of irradiance and angle of incidence, respectively. In addition, $G_f(\theta_{rx})$ denotes the gain of the user's optical filter and $m = -1/\log_2(\cos(\Psi_{1/2}))$, where $\Psi_{1/2}$ is the semi-angle at half power of the LED. Defining $\Psi_{\mathrm{fov}}$ as the half of the PD's FOV, the gain of the optical concentrator $G_c(\theta_{rx})$ is expressed as

$$G_c(\theta_{rx}) = \begin{cases} \dfrac{n_c^2}{\sin^2(\Psi_{\mathrm{fov}})} & \text{if } 0 \le \theta_{rx} \le \Psi_{\mathrm{fov}} \\ 0 & \text{otherwise,} \end{cases}$$

where $n_c$ accounts for the reflective index of the optical concentrator. In outdoor settings, due to the higher elevation of optical BSs and absence of blocking obstacles, the LOS link blockage probability is small. Therefore, we will focus on the case without the LOS blockages, and the scenario with LOS blockages remains open for future studies.

When a user communicates with an optical BS, the user is associated with the optical BS that provides the highest receive power, i.e., the nearest optical BS. Although the user can always be connected to an optical BS when $\Psi_{\mathrm{fov}} = 90°$, assuming that $\Psi_{\mathrm{fov}} = 90°$ is not valid in practical scenarios [24, 25]. When $\Psi_{\mathrm{fov}} < 90°$, the maximum horizontal distance between a user and its serving optical BS is

$$r_{\max} = L\tan(\Psi_{\mathrm{fov}}), \tag{2.6}$$

and hence the user can communicate with an optical BS only when the distance between the user and its closest optical BS is less than $r_{\max}$.

---

[2]When a random PD's orientation is considered [25], one can generalize our analytical results in Sect. 2.3 by computing the expectation of the outage probability introduced in this chapter with respect to the distribution of the random orientation. As the resulting expression has a complicated form, developing a new approach to analyze and optimize VLC networks with random orientation represents an interesting work for future.

Denoting the horizontal location of the nearest optical BS as $\mathbf{x}_V$, we represent the SINR of the VLC networks $\gamma_V$ as [22, 23]

$$\gamma_V = \frac{P_V \kappa^2 (\|\mathbf{x}_V\|^2 + L^2)^{-(m+3)}}{\sum_{\mathbf{x} \in \{\Phi_V \cap \mathcal{B}_o(r_{\max})\} \backslash \mathbf{x}_V} P_V \kappa^2 (\|\mathbf{x}\|^2 + L^2)^{-(m+3)} + \sigma_V^2}$$
$$= \frac{(\|\mathbf{x}_V\|^2 + L^2)^{-(m+3)}}{I_V + \bar{\sigma}_V^2}, \tag{2.7}$$

where $\mathcal{B}_o(r)$ is the ball centered at the origin with radius $r$, $\sigma_V^2$ is the power of noise,[3] $\bar{\sigma}_V^2 \triangleq \sigma_V^2 / (P_V \kappa^2)$ and

$$\mathbf{x}_V = \underset{\mathbf{x} \in \Phi_V \cap \mathcal{B}_o(r_{\max})}{\arg \min} \|\mathbf{x}\|,$$

$$I_V \triangleq \sum_{\mathbf{x} \in \{\Phi_V \cap \mathcal{B}_o(r_{\max})\} \backslash \mathbf{x}_V} (\|\mathbf{x}\|^2 + L^2)^{-(m+3)}$$

$$= \sum_{\mathbf{x} \in \Phi_V} (\|\mathbf{x}\|^2 + L^2)^{-(m+3)} \mathbb{1} \left( \|\mathbf{x}_V\| < \|\mathbf{x}\| \leq r_{\max} \right). \tag{2.8}$$

Note that as $\Psi_{\text{fov}} < 90°$, the signals radiated by the optical BSs located farther than $r_{\max}$ have no impact on the SINR $\gamma_V$.

Since an outage occurs when there is no optical BS within $r_{\max}$ or the SINR $\gamma_V$ in (2.7) is smaller than $\gamma_{\text{th}}$, we can express the outage probability of the VLC networks $\mathcal{P}_o^V(\lambda_V)$ as

$$\mathcal{P}_o^V(\lambda_V) = \mathbb{P}\left(\|\mathbf{x}_V\| > r_{\max}\right) + \mathbb{P}\left(\|\mathbf{x}_V\| \leq r_{\max}, \ \gamma_V < \gamma_{\text{th}}\right)$$
$$= \exp(-\pi \lambda_V r_{\max}^2) + \mathbb{P}\left(\|\mathbf{x}_V\| \leq r_{\max}, \ \gamma_V < \gamma_{\text{th}}\right), \tag{2.9}$$

where $\mathbb{P}\left(\|\mathbf{x}_V\| > r_{\max}\right) = \exp(-\pi \lambda_V r_{\max}^2)$ [6] and the PDF of $\|\mathbf{x}_V\|$ is given by

$$f_{\|\mathbf{x}_V\|}(u) = 2\pi \lambda_V u \exp(-\pi \lambda_V u^2). \tag{2.10}$$

### 2.2.4  Problem Formulation

We assume that users select either an RF network or a VLC network (i.e., multi-modal connection). More specifically, each user first tries to associate with the VLC network (or RF network) and switches to the RF network (or VLC network) if the VLC network (or RF network) is in outage. Therefore, in the RF-optical HetNet,

---

[3]In outdoor scenarios, severe weather conditions and sunlight may cause interference [37]. In this chapter, it is assumed that such interference is treated as noise and is reflected in $\sigma_V^2$.

the outage happens if and only if both RF network and VLC network are in outage. Figure 2.2 illustrates operation scenarios within RF-optical HetNets where each user first attempts to connect to an optical BS. Since RF and VLC links do not interfere with each other, the outage probability of the RF-optical HetNet $\mathcal{P}_o$ is expressed as:

$$\mathcal{P}_o(\lambda_V, \xi) = \mathcal{P}_o^V(\lambda_V)\mathcal{P}_o^R(\xi), \tag{2.11}$$

where $\xi = \lambda_M P_M^{2/\alpha} + \lambda_S P_S^{2/\alpha}$, and $\mathcal{P}_o^V(\lambda_V)$ and $\mathcal{P}_o^R(\xi)$ are defined in (2.9) and (2.3), respectively.

Our objective is to identify the densities $\lambda_V$, $\lambda_M$, and $\lambda_S$ that minimize the APC in (2.1) subject to a constraint on the outage probability in (2.11). Hence, the joint optimal planning problem of the RF-optical HetNet is formulated as follows:

$$\min_{\lambda_V, \lambda_M, \lambda_S} \quad \lambda_V P_V + \lambda_M \bar{P}_M + \lambda_S \bar{P}_S \tag{2.12}$$

$$\text{s.t.} \quad \mathcal{P}_o^V(\lambda_V)\mathcal{P}_o^R(\xi) \leq \epsilon$$

$$0 \leq \lambda_V \leq \lambda_{V,ub}$$

$$0 \leq \lambda_M \leq \lambda_{M,ub}$$

$$0 \leq \lambda_S \leq \lambda_{S,ub},$$

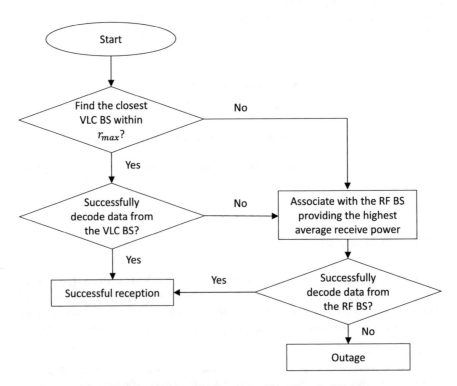

**Fig. 2.2** Scenarios of outage and successful reception within RF-optical HetNet

where $\epsilon$ is a target outage probability, and $\lambda_{V,ub}$, $\lambda_{M,ub}$ and $\lambda_{S,ub}$ are respectively upper bounds on $\lambda_V$, $\lambda_M$ and $\lambda_S$. Here, the upper bounds on the densities can represent the maximum available network deployment cost since the cost increases as the number of BSs gets larger.

In the following section, we derive approximations of $\mathcal{P}_o^V(\lambda_V)$ in (2.9), which lead to approximations of the outage probability of RF-optical HetNets $\mathcal{P}_o(\lambda_V, \xi)$ in (2.11). Then, based on the approximation, efficient algorithms that optimize the BS densities for VLC networks and RF-optical HetNets are introduced in Sect. 2.4.

## 2.3  Outage Probability of VLC Network

In this section, we examine the outage probability of the VLC network $\mathcal{P}_o^V(\lambda_V)$ in (2.9) for an arbitrary $\Psi_{\text{fov}}$. First, we rewrite $\mathbb{P}\left(\|\mathbf{x}_V\| \le r_{\max}, \; \gamma_V < \gamma_{\text{th}}\right)$ in (2.9) as

$$
\mathbb{P}\left(\|\mathbf{x}_V\| \le r_{\max}, \; \gamma_V < \gamma_{\text{th}}\right) \tag{2.13}
$$
$$
= \mathbb{P}\left(\|\mathbf{x}_V\| \le r_{\max}, \; I_V > \frac{(\|\mathbf{x}_V\|^2 + L^2)^{-(m+3)}}{\gamma_{\text{th}}} - \bar{\sigma}_V^2\right)
$$
$$
= \int_0^{r_{\max}} \mathbb{P}\left(I_V > \frac{(u^2 + L^2)^{-(m+3)}}{\gamma_{\text{th}}} - \bar{\sigma}_V^2 \;\Big|\; u\right) f_{\|\mathbf{x}_V\|}(u)\,du,
$$

where $f_{\|\mathbf{x}_V\|}(u)$ stands for the PDF of the distance between a user and its serving optical BS and assumes the expression in (2.10).

It should be noted that in order to find an expression of $\mathcal{P}_o^V(\lambda_V)$ in (2.9), we need to derive an analytical representation for the complimentary cumulative distribution function (CCDF) of $I_V$ in (2.13). However, it is difficult to identify an exact expression for the CCDF of $I_V$ with an arbitrary $\Psi_{\text{fov}}$ since it is intractable to obtain a closed-form representation of the inverse Laplace transform of $I_V$ [23].

Define $\tilde{I}_V \triangleq \sum_{\mathbf{x} \in \Phi_V} (\|\mathbf{x}\|^2 + L^2)^{-(m+3)} \mathbb{1}\left(\|\mathbf{x}_V\| < \|\mathbf{x}\|\right)$ as $I_V$ in (2.8) with $\Psi_{\text{fov}} = 90°$, i.e., $r_{\max} \to \infty$. In [21], an approximation of the CCDF of $\tilde{I}_V$ was introduced by approximating the PDF of $\tilde{I}_V$ as a sum of Gamma distributions. By adopting the result in [21], an approximation of the CCDF of $\tilde{I}_V$ is expressed as

$$
\mathbb{P}\left(\tilde{I}_V > \frac{(u^2 + L^2)^{-(m+3)}}{\gamma_{\text{th}}} - \bar{\sigma}_V^2 \;\Big|\; u\right) \tag{2.14}
$$
$$
\approx \sum_{k=0}^{\infty} \frac{\Delta_k^{(1)}(\lambda_V)\Delta_k^{(2)}(\lambda_V)}{k!\,\Gamma(k + \varphi(\lambda_V))} \approx \sum_{k=0}^{K_{\max}} \frac{\Delta_k^{(1)}(\lambda_V)\Delta_k^{(2)}(\lambda_V)}{k!\,\Gamma(k + \varphi(\lambda_V))},
$$

where $K_{\max}$ denotes a truncation parameter and

$$\Delta_k^{(1)}(\lambda_V) = \beta^{\varphi(\lambda_V)} \sum_{i=0}^{k} C_i^{(k)} \mu_i,$$

$$\Delta_k^{(2)}(\lambda_V) = \sum_{j=0}^{k} \frac{1}{\beta^{j+\varphi(\lambda_V)}} C_j^{(k)} \Gamma\big(j+\varphi(\lambda_V), \beta(Q/\gamma_{th} - \bar{\sigma}_V^2)\big),$$

$$\varphi(\lambda_V) = \frac{\lambda_V \pi (2m+5)}{(m+2)^2}(u^2 + L^2),$$

$$\beta = \frac{2m+5}{(m+2)}(u^2+L^2)^{m+3}, \quad Q = (u^2+L^2)^{-(m+3)},$$

$$C_l^{(k)} = (-1)^{k-l}\binom{k}{l}\beta^l s_l^{(k)}, \quad \eta_l = \frac{\lambda_V \pi (u^2+L^2)^{1-l(m+3)}}{l(m+3)-1},$$

$$\mu_l = \begin{cases} 1 & \text{if } l=0 \\ \eta_1 & \text{if } l=1 \\ \eta_l + \sum_{q=1}^{l-1}\binom{l-1}{q-1}\eta_q \mu_{l-q} & \text{if } l \geq 2, \end{cases}$$

$$s_l^{(k)} = \begin{cases} 1 & \text{if } l > k-1 \\ \prod_{q=l}^{k-1}(\varphi(\lambda_V)+q) & \text{if } l \leq k-1, \end{cases}$$

where $\Gamma(s) = \int_0^\infty t^{s-1}e^{-t}dt$ and $\Gamma(s,z) = \int_z^\infty t^{s-1}e^{-t}dt$ stand for the Gamma function and the upper incomplete Gamma function, respectively. Note that $\tilde{I}_V$ overestimates the actual interference $I_V$ as $\tilde{I}_V$ includes signals that have no impact on the SINR due to the limited FOV. Hence, by using the CCDF of $\tilde{I}_V$ in (2.14), an upper bound of $\mathcal{P}_o^V(\lambda_V)$ is obtained in the following lemma.

**Lemma 2.1** *An upper bound of the outage probability of the VLC network* $\mathcal{P}_o^V(\lambda_V)$ *in (2.9) is given by*

$$\mathcal{P}_{o,u}^V(\lambda_V) = \exp(-\pi\lambda r_{max}^2) \tag{2.15}$$
$$+ \int_0^{r_{max}} \frac{2\pi\lambda_V u}{\exp(\pi\lambda_V u^2)} \sum_{k=0}^{K_{max}} \frac{\Delta_k^{(1)}(\lambda_V)\Delta_k^{(2)}(\lambda_V)}{k!\Gamma(k+\varphi(\lambda_V))} du.$$

**Proof** The expression in (2.15) is obtained by replacing $I_V$ in (2.13) with $\tilde{I}_V$, and plugging (2.13) and (2.14) into (2.9). □

In order to find a tight approximation for the CCDF of $I_V$ in (2.13), we first define a PPP $\bar{\Phi}_V$ with density $\bar{\lambda}_V$ and $\bar{I}_V \triangleq \sum_{x \in \bar{\Phi}_V}(\|x\|^2 + L^2)^{-(m+3)}\mathbb{1}(\|x_V\| < \|x\|)$, and then calculate a proper value of $\bar{\lambda}_V$ that can reflect the influence of $r_{max}$ in (2.6) on $I_V$, i.e., $\bar{I}_V \approx I_V$. Thus, we compute $\bar{\lambda}_V$ which fulfills $\mathbb{E}[I_V] = \mathbb{E}[\bar{I}_V]$. Employing Campbell's Theorem [6], we obtain [43, 2.147.2]

$$\mathbb{E}[I_V] = 2\pi\lambda_V \int_{\|\mathbf{x}_V\|}^{r_{max}} (x^2 + L^2)^{-(m+3)} x\,dx$$

$$= \frac{2\pi\lambda_V}{2m+4} \left( \left(\|\mathbf{x}_V\|^2 + h^2\right)^{-m-2} - (r_{max}^2 + h^2)^{-m-2} \right),$$

$$\mathbb{E}[\bar{I}_V] = 2\pi\bar{\lambda}_V \int_{\|\mathbf{x}_V\|}^{\infty} (x^2 + L^2)^{-(m+3)} x\,dx$$

$$= \frac{2\pi\bar{\lambda}_V}{2m+4} \left(\|\mathbf{x}_V\|^2 + h^2\right)^{-m-2}.$$

Thus, $\bar{\lambda}_V$ is expressed as

$$\bar{\lambda}_V = \lambda_V \frac{\left(\|\mathbf{x}_V\|^2 + h^2\right)^{-m-2} - (r_{max}^2 + h^2)^{-m-2}}{\left(\|\mathbf{x}_V\|^2 + h^2\right)^{-m-2}}$$

$$= \lambda_V \left( 1 - \frac{\left(\|\mathbf{x}_V\|^2 + L^2\right)^{m+2}}{(r_{max}^2 + L^2)^{m+2}} \right). \tag{2.16}$$

Note that the scaled density $\bar{\lambda}_V$ is determined to match the first moment of the actual interference $I_V$. Thus, by leveraging the scaled density $\bar{\lambda}_V$ in (2.16), we derive approximations of $\mathcal{P}_o^V(\lambda_V)$ in (2.9) as described in the following theorem.

**Theorem 2.1** *An approximation of the outage probability of the VLC network $\mathcal{P}_o^V(\lambda_V)$ in (2.9) is given by*

$$\bar{\mathcal{P}}_o^V(\lambda_V) = \exp(-\pi\lambda_V r_{max}^2) + \int_0^{r_{max}} \frac{2\pi\lambda_V u}{\exp(\pi\lambda_V u^2)} \sum_{k=0}^{K_{max}} \frac{\Delta_k^{(1)}(\bar{\lambda}_V)\Delta_k^{(2)}(\bar{\lambda}_V)}{k!\Gamma(k + \varphi(\bar{\lambda}_V))} du, \tag{2.17}$$

*where $\bar{\lambda}_V$ has been defined in (2.16). Furthermore, by setting the truncation parameter $K_{max}$ to 0 and assuming that the impact of $\bar{\sigma}_V^2$ on $\mathcal{P}_o^V(\lambda_V)$ is negligible ($\bar{\sigma}_V^2 = 0$), a more simplified approximation is expressed as*

$$\hat{\mathcal{P}}_o^V(\lambda_V) = \exp(-\pi\lambda_V r_{max}^2) + \int_0^{r_{max}} \frac{2\pi\lambda_V u}{\exp(\pi\lambda_V u^2)} \frac{\Gamma\left(\varphi(\bar{\lambda}_V), \frac{2m+5}{(m+2)\gamma_{th}}\right)}{\Gamma(\varphi(\bar{\lambda}_V))} du. \tag{2.18}$$

***Proof*** By replacing $\lambda_V$ in (2.14) with $\bar{\lambda}_V$ in (2.16) and substituting (2.13) and (2.14) into (2.9), the result in (2.17) is obtained. Also, setting up $K_{max} = 0$ and $\bar{\sigma}_V^2 = 0$, we have

$$\Delta_0^{(1)}(\bar{\lambda}_V)\Delta_0^{(2)}(\bar{\lambda}_V) = \beta^{\varphi(\bar{\lambda}_V)} \frac{1}{\beta^{\varphi(\bar{\lambda}_V)}} \Gamma\left(\varphi(\bar{\lambda}_V), \frac{\beta Q}{\gamma_{\text{th}}}\right)$$

$$= \Gamma\left(\varphi(\bar{\lambda}_V), \frac{2m+5}{(m+2)\gamma_{\text{th}}}\right),$$

which leads to the expression in (2.18).                                                                  □

Note that the approximation in (2.17) contains only a single integral of a summation term, and the summation term is further simplified in (2.18). However, even with the expression (2.18), due to the Gamma function, upper incomplete Gamma function, and integral operation, it is difficult to mathematically identify the impact of $\lambda_V$ on the outage probability. Therefore, instead using a mathematical approach, we corroborate that the outage probability of the VLC network is either a decreasing function or a convex function with respect to $\lambda_V$ based on the numerical simulations shown in Figs. 2.3 and 2.4. In Sect. 2.4, the optimization problem in (2.12) will be solved based on the approximation (2.18) and the observation offered by the numerical simulations.

Figures 2.3 and 2.4 depict the outage probability of VLC networks $\mathcal{P}_o^V(\lambda_V)$ for the network parameters listed in Table 2.1. First of all, it is shown that our approximation

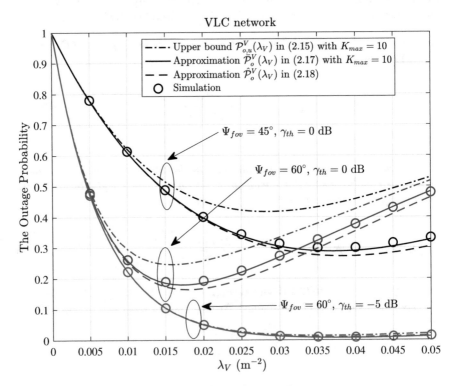

**Fig. 2.3** Outage probability of VLC networks as a function of $\lambda_V$

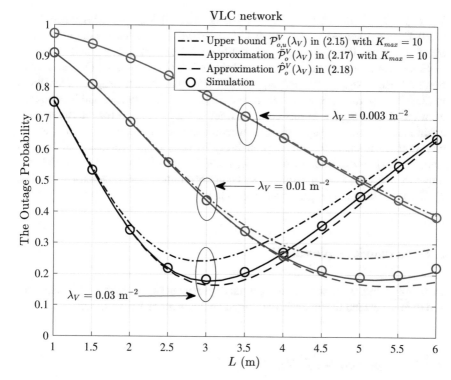

**Fig. 2.4** Outage probability of VLC networks as a function of $L$

**Table 2.1** VLC network parameters

| Symbol | $P_V$ | $\Psi_{1/2}$ | $A_{pd}$ | $\eta$ | $G_f(\theta_{rx})$ | $n_c$ | $\sigma_V^2$ |
|---|---|---|---|---|---|---|---|
| Value | 1 W | 60° | $10^{-4}$ m$^2$ | 0.4 A/m | 1 | 1.5 | −117 dBm |

in (2.17) with $K_{\max} = 10$ predicts well the actual outage probability and the approximation in (2.18) exhibits only small gaps compared to the Monte Carlo simulation results for various values of $\lambda_V$, $\Psi_{fov}$ and $\gamma_{th}$. In addition, we can see that the upper bound in (2.15) which is based on the assumption that $\Psi_{fov} = 90°$ experiences large gaps compared to the Monte Carlo simulation results.

In Fig. 2.3, we evaluate $\mathcal{P}_o^V(\lambda_V)$ as a function of $\lambda_V$ when $L = 4$ m. From the figure, we observe that $\mathcal{P}_o^V(\lambda_V)$ is either a decreasing or a convex function with respect to $\lambda_V$. More specifically, when $\lambda_V$ is small, $\mathcal{P}_o^V(\lambda_V)$ decays as $\lambda_V$ increases due to the term $\mathbb{P}(\|\mathbf{x}_V\| > r_{\max}) = \exp(-\pi \lambda_V r_{\max}^2)$ in (2.9) which reflects the probability that there is no optical BS within the distance $r_{\max}$. In other words, $\mathbb{P}(\|\mathbf{x}_V\| > r_{\max})$ in (2.9) dominates $\mathcal{P}_o^V(\lambda_V)$ when $\lambda_V$ is small. If the impact of interference on $\mathcal{P}_o^V(\lambda_V)$ is more pronounced than that of $\mathbb{P}(\|\mathbf{x}_V\| > r_{\max})$ and the desired term, $\mathcal{P}_o^V(\lambda_V)$ grows as $\lambda_V$ becomes bigger when $\lambda_V$ is large. Since a lower $\Psi_{fov}$ incurs a higher

$\mathbb{P}(\|\mathbf{x}_V\| > r_{\max})$, the outage probability gets larger as $\Psi_{\text{fov}}$ degrades when $\lambda_V$ is small. On the other hand, when $\lambda_V$ is high, a growth of $\Psi_{\text{fov}}$ leads to an increase of interference, and thus the outage probability decreases as $\Psi_{\text{fov}}$ gets smaller.

Figure 2.4 examines the outage probability of VLC networks $\mathcal{P}_o^V(\lambda_V)$ as a function of $L$ when $\Psi_{\text{fov}} = 60°$ and $\gamma_{\text{th}} = 0$ dB. Note that the term $r_{\max} = L \tan(\Psi_{\text{fov}})$ in (2.6) is proportional to $L$, and therefore $\mathbb{P}(\|\mathbf{x}_V\| > r_{\max})$ decays and interference grows as $L$ gets bigger. In this context, when $\mathbb{P}(\|\mathbf{x}_V\| > r_{\max})$ (or interference) is a dominant factor, $\mathcal{P}_o^V(\lambda_V)$ degrades (or increases) as $L$ becomes larger. For example, when $\lambda_V$ is small, the outage probability $\mathcal{P}_o^V(\lambda_V)$ keeps decreasing as $L$ grows since $\mathbb{P}(\|\mathbf{x}_V\| > r_{\max})$ dominates $\mathcal{P}_o^V(\lambda_V)$. On the other hand, when $\lambda_V$ is high, the influence of $\mathbb{P}(\|\mathbf{x}_V\| > r_{\max})$ on $\mathcal{P}_o^V(\lambda_V)$ quickly diminishes as $L$ increases, and thus $\mathcal{P}_o^V(\lambda_V)$ gets bigger as $L$ grows when $L$ is large.

## 2.4   Network Deployment

In the following, we first examine optimal deployment of standalone VLC networks and then we extend the analysis to optimal deployment of RF-VLC HetNets.

### 2.4.1   Deployment of VLC Network

In this subsection, we investigate the optimization of the optical BS density $\lambda_V$ that minimizes the APC under a constraint on the outage probability. From Fig. 2.3, it is shown that the approximation $\hat{\mathcal{P}}_o^V(\lambda_V)$ in (2.18) predicts well the outage probability with a low computational complexity. Hence, we formulate the APC minimization problem for VLC networks by using $\hat{\mathcal{P}}_o^V(\lambda_V)$ as a constraint:

$$\min_{\lambda_V} \ \lambda_V P_V \tag{2.19}$$

$$\text{s.t.} \ \ \hat{\mathcal{P}}_o^V(\lambda_V) \leq \epsilon$$

$$0 \leq \lambda_V \leq \lambda_{V,ub}.$$

From Fig. 2.3 it turns out that depending on network parameters, $\hat{\mathcal{P}}_o^V(\lambda_V)$ is either a convex or a decreasing function of $\lambda_V$. Let us denote $g(x)$ as the derivative of $\hat{\mathcal{P}}_o^V(x)$ with respect to $x$, i.e., $g(x) \triangleq \frac{\partial \hat{\mathcal{P}}_o^V(x)}{\partial x} = \lim_{\delta \to 0} \frac{\hat{\mathcal{P}}_o^V(x+\delta) - \hat{\mathcal{P}}_o^V(x)}{\delta}$. Then, the cases with $g(\lambda_{V,ub}) < 0$ and $g(\lambda_{V,ub}) \geq 0$ respectively mean that $\hat{\mathcal{P}}_o^V(\lambda_V)$ is a decreasing and a convex function with $\lambda_V \in [0, \ \lambda_{V,ub}]$. When $g(\lambda_{V,ub}) \geq 0$, i.e., $\hat{\mathcal{P}}_o^V(\lambda_V)$ is a convex function with respect to $\lambda_V$, one can determine

$$\lambda_{V,min} \triangleq \underset{\lambda_V \in [0, \ \lambda_{V,ub}]}{\arg \min} \ \hat{\mathcal{P}}_o^V(\lambda_V), \tag{2.20}$$

by exploiting the *golden section search method* [44]. When compared to $\lambda_{V,min}$, any $\lambda_V$ in $(\lambda_{V,min}, \ \lambda_{V,ub}]$ leads to increased outage probability and APC, and therefore, the optimal $\lambda_V$ for the problem in (2.19) should be in $[0, \ \lambda_{V,min}]$. Let us define $\tilde{\lambda}_{V,ub}$ as

$$\tilde{\lambda}_{V,ub} = \begin{cases} \lambda_{V,ub} & \text{if } g(\lambda_{V,ub}) < 0 \\ \lambda_{V,min} & \text{otherwise.} \end{cases} \tag{2.21}$$

Then, since $\hat{\mathcal{P}}_o^V(\lambda_V)$ is always a decreasing function of $\lambda_V \in [0, \ \tilde{\lambda}_{V,ub}]$, the optimal $\lambda_V$ must satisfy the constraint on the outage probability with equality, i.e., $\hat{\mathcal{P}}_o^V(\lambda_V) = \epsilon$. Thus, the problem in (2.19) is reformulated as

$$\min_{\lambda_V} \ \lambda_V \tag{2.22}$$
$$\text{s.t. } \hat{\mathcal{P}}_o^V(\lambda_V) = \epsilon$$
$$0 \leq \lambda_V \leq \tilde{\lambda}_{V,ub},$$

and the optimal solution can be identified by leveraging *the bisection algorithm* [44] if $\hat{\mathcal{P}}_o^V(\tilde{\lambda}_{V,ub}) \leq \epsilon$. The problem is not feasible if $\hat{\mathcal{P}}_o^V(\tilde{\lambda}_{V,ub}) > \epsilon$. The algorithm of the introduced method is summarized as follows:

---

Algorithm 1. Optical BS Density Optimization

---

1. Compute $g(\lambda_{V,ub})$ and $\lambda_{V,min}$ in (2.20) using the golden section search method when $g(\lambda_{V,ub}) \geq 0$.
2. Calculate $\tilde{\lambda}_{V,ub}$ in (2,21).
3. Solve the problem in (2,22) using the bisection algorithm.

---

An important point to note here is that Algorithm 1 is based on the simple expression of the approximation of the outage probability in (2.18). In addition, Algorithm 1 employs efficient search algorithms, e.g., golden section search method and bisection algorithm. The numbers of iterations for the golden section search and bisection algorithms are respectively $\log_{1/0.618}(\Omega/\tau)$ and $\log_2(\Omega/\tau)$, where $\Omega$ and $\tau$ denote the length of the search range and an acceptable tolerance, respectively. Hence, Algorithm 1 presents a low computational complexity. From the simulation results to be introduced in Sect. 2.5, it will be confirmed that Algorithm 1 exhibits almost identical performance to an algorithm that exhaustively searches the optimal value of $\lambda_V$ exploiting the approximation in (2.17), which is very accurate but has a higher complexity compared to that of the approximation in (2.18).

### *2.4.2   Deployment of RF-Optical HetNet*

We now focus on optimal deployment of RF-optical HetNets as described in (2.12) by leveraging the approximation of the outage probability of VLC networks $\hat{\mathcal{P}}_o^V(\lambda_V)$ in (2.18). As addressed in Sect. 2.4.1, the optimal $\lambda_V$ should be in $[0, \tilde{\lambda}_{V,ub}]$ and $\hat{\mathcal{P}}_o^V(\lambda_V)$ decays as $\lambda_V \in [0, \tilde{\lambda}_{V,ub}]$ gets larger, and $\tilde{\lambda}_{V,ub}$ is defined in (2.21). Also, the outage probability of RF networks $\mathcal{P}_o^R(\xi)$ in (2.3) is a non-increasing function of $\xi = \lambda_M P_M^{2/\alpha} + \lambda_S P_S^{2/\alpha}$. Since the APC in (2.1) increases with $\lambda_V$, $\lambda_M$ and $\lambda_S$, the constraint on the outage probability in (2.12) should be met with equality, and therefore the problem can be formulated as

$$\min_{\lambda_V, \lambda_M, \lambda_S} \quad \lambda_V P_V + \lambda_M \bar{P}_M + \lambda_S \bar{P}_S \tag{2.23}$$

$$\text{s.t. } \hat{\mathcal{P}}_o^V(\lambda_V) \mathcal{P}_o^R(\xi) = \epsilon$$

$$0 \le \lambda_V \le \tilde{\lambda}_{V,ub}$$

$$0 \le \lambda_M \le \lambda_{M,ub}$$

$$0 \le \lambda_S \le \lambda_{S,ub}.$$

Let us first investigate $\mathcal{P}_o^R(\xi)$. As $\mathcal{P}_o^R(\xi)$ is a non-increasing function of $\xi$, any $\xi$ such that $\mathcal{P}_o^R(\xi) < \epsilon$ cannot be the optimal solution as it results in an increase of the objective function in (2.23). From this observation, we obtain a condition on $\lambda_M$ and $\lambda_S$ as

$$0 \le \lambda_M P_M^{2/\alpha} + \lambda_S P_S^{2/\alpha} \le \tilde{\xi}_{ub},$$

where $\xi_{ub} \triangleq \lambda_{M,ub} P_M^{2/\alpha} + \lambda_{S,ub} P_S^{2/\alpha}$ and

$$\tilde{\xi}_{ub} = \begin{cases} \xi \text{ such that } \mathcal{P}_o^R(\xi) = \epsilon & \text{if } \mathcal{P}_o^R(\xi_{ub}) < \epsilon \\ \xi_{ub} & \text{otherwise.} \end{cases} \tag{2.24}$$

We define $\mathcal{P}_{o,lb}^R$ as a lower bound of $\mathcal{P}_o^R(\xi)$ which is equal to $\mathcal{P}_{o,lb}^R = \mathcal{P}_o^R(\tilde{\xi}_{ub})$. Then, since $\mathcal{P}_{o,lb}^R \le \mathcal{P}_o^R(\xi) \le 1$ for $\xi \in [0, \tilde{\xi}_{ub}]$ and $\hat{\mathcal{P}}_o^V(\lambda_V)\mathcal{P}_o^R(\xi) = \epsilon$, we have

$$\mathcal{P}_{o,lb}^R \le \frac{\epsilon}{\hat{\mathcal{P}}_o^V(\lambda_V)} \le 1 \quad \Leftrightarrow \quad \epsilon \le \hat{\mathcal{P}}_o^V(\lambda_V) \le \frac{\epsilon}{\mathcal{P}_{o,lb}^R}$$

$$\Leftrightarrow \quad \hat{\lambda}_{V,lb} \le \lambda_V \le \hat{\lambda}_{V,ub}, \tag{2.25}$$

where $\hat{\lambda}_{V,lb}$ is the value satisfying $\hat{\mathcal{P}}_o^V(\hat{\lambda}_{V,lb}) = \epsilon/\mathcal{P}_{o,lb}^R$ and

$$\hat{\lambda}_{V,ub} = \begin{cases} \lambda \text{ such that } \hat{\mathcal{P}}_o^V(\lambda) = \epsilon & \text{if } \hat{\mathcal{P}}_o^V(\tilde{\lambda}_{V,ub}) < \epsilon, \\ \tilde{\lambda}_{V,ub} & \text{otherwise.} \end{cases} \tag{2.26}$$

Here, one can readily get $\tilde{\xi}_{ub}$ in (2.24), $\hat{\lambda}_{V,lb}$ and $\hat{\lambda}_{V,ub}$ in (2.26) by utilizing the bisection search method as both $\mathcal{P}_o^R(\xi)$ and $\hat{\mathcal{P}}_o^V(\lambda_V)$ are decreasing functions of $\xi \in [0, \xi_{ub}]$ and $\lambda_V \in [0, \tilde{\lambda}_{V,ub}]$, respectively.

Let us denote $\hat{\xi}(\lambda_V)$ as $\xi$ that fulfills $\hat{\mathcal{P}}_o^V(\lambda_V)\mathcal{P}_o^R(\xi) = \epsilon$ for a given $\lambda_V$. Then, we can reformulate the problem in (2.23) as

$$\min_{\lambda_V, \lambda_M, \lambda_S} \quad \lambda_V P_V + \lambda_M \bar{P}_M + \lambda_S \bar{P}_S \tag{2.27}$$

$$\text{s.t. } \lambda_M P_M^{2/\alpha} + \lambda_S P_S^{2/\alpha} = \hat{\xi}(\lambda_V)$$

$$\hat{\lambda}_{V,lb} \leq \lambda_V \leq \hat{\lambda}_{V,ub}$$

$$0 \leq \lambda_M \leq \lambda_{M,ub}$$

$$0 \leq \lambda_S \leq \lambda_{S,ub}.$$

From the equality $\lambda_M P_M^{2/\alpha} + \lambda_S P_S^{2/\alpha} = \hat{\xi}(\lambda_V)$ in (2.27), we get

$$\lambda_S = P_S^{-2/\alpha}\left(\hat{\xi}(\lambda_V) - \lambda_M P_M^{2/\alpha}\right), \tag{2.28}$$

and thus the region of $\lambda_M$ can be found as

$$0 \leq \lambda_S \leq \lambda_{S,ub} \tag{2.29}$$

$$\Leftrightarrow \hat{\xi}(\lambda_V) - \lambda_{S,ub} P_S^{2/\alpha} \leq \lambda_M P_M^{2/\alpha} \leq \hat{\xi}(\lambda_V)$$

$$\Leftrightarrow \tilde{\lambda}_{M,lb}(\lambda_V) \leq \lambda_M \leq \tilde{\lambda}_{M,ub}(\lambda_V),$$

where $\tilde{\lambda}_{M,lb}(\lambda_V) \triangleq P_M^{-2/\alpha}\left(\hat{\xi}(\lambda_V) - \lambda_{S,ub} P_S^{2/\alpha}\right)$ and $\tilde{\lambda}_{M,ub}(\lambda_V) \triangleq P_M^{-2/\alpha}\hat{\xi}(\lambda_V)$. Also, using (2.28), we rewrite $\lambda_M \bar{P}_M + \lambda_S \bar{P}_S$ in the objective function of the problem (2.27) as

$$\lambda_M \bar{P}_M + \lambda_S \bar{P}_S = \lambda_M \bar{P}_M + P_S^{-2/\alpha}\left(\hat{\xi}(\lambda_V) - \lambda_M P_M^{2/\alpha}\right)\bar{P}_S$$

$$= \lambda_M \left(\bar{P}_M - P_S^{-2/\alpha} P_M^{2/\alpha} \bar{P}_S\right) + \hat{\xi}(\lambda_V) P_S^{-2/\alpha} \bar{P}_S$$

$$= \lambda_M C_1 + \hat{\xi}(\lambda_V) C_2, \tag{2.30}$$

where $C_1 \triangleq \bar{P}_M - P_S^{-2/\alpha} P_M^{2/\alpha} \bar{P}_S$ and $C_2 \triangleq P_S^{-2/\alpha} \bar{P}_S$.

By plugging (2.29) and (2.30) into (2.27), the problem in (2.27) is simplified to *a two-variable problem* as follows:

$$\min_{\lambda_V, \lambda_M} \ \lambda_V P_V + \lambda_M C_1 + \hat{\xi}(\lambda_V)C_2 \tag{2.31}$$

$$\text{s.t. } \hat{\lambda}_{V,lb} \le \lambda_V \le \hat{\lambda}_{V,ub}$$

$$\hat{\lambda}_{M,lb}(\lambda_V) \le \lambda_M \le \hat{\lambda}_{M,ub}(\lambda_V),$$

where $\hat{\lambda}_{M,lb}(\lambda_V) = \max(\tilde{\lambda}_{M,lb}(\lambda_V), 0)$ and $\hat{\lambda}_{M,ub}(\lambda_V) = \min(\tilde{\lambda}_{M,ub}(\lambda_V), \lambda_{M,ub})$. Here, for a given $\lambda_V$, the optimal $\lambda_M$ can be expressed as

$$\lambda_M(\lambda_V) = \begin{cases} \hat{\lambda}_{M,lb}(\lambda_V) & \text{if } C_1 \ge 0 \\ \hat{\lambda}_{M,ub}(\lambda_V) & \text{otherwise.} \end{cases} \tag{2.32}$$

Hence, by combining (2.31) and (2.32), the problem in (2.31) can be further simplified to *a single-variable problem* as follows:

$$\min_{\lambda_V} \ \lambda_V P_V + \lambda_M(\lambda_V)C_1 + \hat{\xi}(\lambda_V)C_2 \tag{2.33}$$

$$\text{s.t. } \hat{\lambda}_{V,lb} \le \lambda_V \le \hat{\lambda}_{V,ub}.$$

Finally, by exploiting one-dimensional line search algorithms, we can identify the optimal value of $\lambda_V$ and the corresponding $\lambda_M$ and $\lambda_S$ from (2.32) and (2.28), respectively. We summarize our introduced algorithm to obtain $\lambda_V$, $\lambda_M$ and $\lambda_S$ as follows:

---

**Algorithm 2. Optical BSs, MBSs and SBSs Densities Optimization**

---

1. Compute $\tilde{\lambda}_{V,ub}$ in (2.21) and $\tilde{\xi}_{ub}$ in (2.24).
2. Determine $\hat{\lambda}_{V,lb}$ and $\hat{\lambda}_{V,ub}$ by using (2.25) and (2.26).
3. Identify $\lambda_V$ by solving the problem in (2.33).
4. Obtain $\lambda_M$ by utilizing (2.32).
5. Calculate $\lambda_S$ by leveraging (2.28).

---

As a special case, let us look at an RF-optical HetNet with a single-tier of RF BSs, i.e., MBSs. In this case, from the result in (2.3), the outage probability of the RF networks is expressed as

$$\mathcal{P}_o^R(\lambda_M) = 1 - \pi \lambda_M P_M^{2/\alpha} \int_0^\infty \exp\left(-\gamma_{\text{th}}\sigma_R^2 t^{\alpha/2}\right)$$

$$\times \exp\left(-\pi t(1 + \rho(\gamma_{\text{th}}, \alpha))\lambda_M P_M^{2/\alpha}\right) dt.$$

Since $\mathcal{P}_o^R(\lambda_M)$ is a non-increasing function with $\lambda_M$, if $\mathcal{P}_o^R(\lambda_{M,ub}) < \epsilon$, one could find $\lambda$ such that $\mathcal{P}_o^R(\lambda) = \epsilon$, and $\lambda_M \in [\lambda, \lambda_{M,ub}]$ cannot be the optimal solution as it incurs an increase of the APC. In addition, from Sect. 2.4.1, it is shown that $\hat{\mathcal{P}}_o^V(\lambda_V)$ is a decreasing function with $\lambda_V \in [0, \tilde{\lambda}_{V,ub}]$ where $\tilde{\lambda}_{V,ub}$ is defined in (2.21). Therefore, we formulate the optimization problem as

$$\min_{\lambda_V, \lambda_M} \lambda_V P_V + \lambda_M \bar{P}_M \tag{2.34}$$

$$\text{s.t.} \ \hat{\mathcal{P}}_o^V(\lambda_V)\mathcal{P}_o^R(\lambda_M) = \epsilon$$

$$0 \le \lambda_V \le \tilde{\lambda}_{V,ub}$$

$$0 \le \lambda_M \le \tilde{\lambda}_{M,ub},$$

where $\tilde{\lambda}_{M,ub}$ is given by

$$\tilde{\lambda}_{M,ub} = \begin{cases} \lambda \text{ such that } \mathcal{P}_o^R(\lambda) = \epsilon & \text{if } \mathcal{P}_o^R(\lambda_{M,ub}) < \epsilon \\ \lambda_{M,ub} & \text{otherwise.} \end{cases} \tag{2.35}$$

Let us denote $\hat{\lambda}_M(\lambda_V)$ as $\lambda_M$ which meets $\mathcal{P}_o^R(\lambda_M) = \epsilon/\hat{\mathcal{P}}_o^V(\lambda_V)$ for a given $\lambda_V$. As $\mathcal{P}_o^R(\tilde{\lambda}_{M,ub}) \le \mathcal{P}_o^R(\lambda_M) \le 1$, we have the range of $\lambda_V$ as

$$\mathcal{P}_o^R(\tilde{\lambda}_{M,ub}) \le \frac{\epsilon}{\hat{\mathcal{P}}_o^V(\lambda_V)} \le 1$$

$$\Leftrightarrow \epsilon \le \hat{\mathcal{P}}_o^V(\lambda_V) \le \frac{\epsilon}{\mathcal{P}_o^R(\tilde{\lambda}_{M,ub})}$$

$$\Leftrightarrow \hat{\lambda}_{V,lb} \le \lambda_V \le \hat{\lambda}_{V,ub}, \tag{2.36}$$

where $\hat{\lambda}_{V,lb}$ is the value satisfying $\hat{\mathcal{P}}_o^V(\hat{\lambda}_{V,lb}) = \epsilon/\mathcal{P}_o^R(\tilde{\lambda}_{M,ub})$ and $\hat{\lambda}_{V,ub}$ is defined in (2.26). Then, by combining (2.34) and (2.36), the problem in (2.34) is simplified to a *single-variable problem* as

$$\min_{\lambda_V} \lambda_V P_V + \hat{\lambda}_M(\lambda_V)\bar{P}_M \tag{2.37}$$

$$\text{s.t.} \ \hat{\lambda}_{V,lb} \le \lambda_V \le \hat{\lambda}_{V,ub}.$$

Lastly, by using one-dimensional line search algorithms, we can obtain the optimal value of $\lambda_V$ and the corresponding $\lambda_M$. The introduced algorithm for the RF-optical HetNets with a single-tier of RF BSs is summarized as follows:

---

Algorithm 3. Optical BSs and MBSs Densities Optimization

---

1. Compute $\tilde{\lambda}_{V,ub}$ in (2.21) and $\tilde{\lambda}_{M,ub}$ in (2.35).
2. Determine $\hat{\lambda}_{V,lb}$ and $\hat{\lambda}_{V,ub}$ in (2.36).
3. Identify $\lambda_V$ by solving the problem in (2.37).
4. Find $\lambda_M$ which fulfills $\mathcal{P}_o^R(\lambda_M) = \epsilon/\hat{\mathcal{P}}_o^V(\lambda_V)$.

---

From the numerical results in Sect. 2.5, it will be verified that Algorithms 2 and 3 exhibit negligible performance loss compared to the exhaustive algorithms which find the optimal values of $\lambda_M$, $\lambda_S$ and $\lambda_V$ by comparing all combinations of $\lambda_M \in [0, \lambda_{M,ub}]$, $\lambda_S \in [0, \lambda_{S,ub}]$ and $\lambda_V \in [0, \lambda_{V,ub}]$.

## 2.5  Simulation Results

In this section, we provide simulation results to show the effectiveness of RF-optical HetNets and validate the efficiency of the introduced algorithms to develop energy efficient RF-optical HetNets. We set $\lambda_{V,ub} = 0.1$ m$^{-2}$, $\lambda_{M,ub} = 10^{-5}$ m$^{-2}$ and $\lambda_{S,ub} = 10^{-6}$ m$^{-2}$, and use the VLC network parameters in Tables 2.1 and the RF network parameters in Table 2.2.

Figure 2.5 demonstrates the APC of VLC networks with Algorithm 1 and an exhaustive search algorithm that manually searches the optimal density of optical BSs $\lambda_V$ when utilizing the approximation in (2.17). First, we confirm that Algorithm 1 exhibits almost identical performance to the exhaustive search algorithm with much reduced computational complexity. As expected, the APC of the VLC networks $P(\lambda_V)$ decays as the target outage probability $\epsilon$ becomes larger and $\gamma_{th}$ decreases. Also, it is seen that the APC increases when $L$ or $\Psi_{fov}$ gets smaller. Note that if $\gamma_{th}$ is high, the VLC networks tend to fail in achieving small values of target outage probability $\epsilon$, and hence the minimum achievable $\epsilon$ grows with $\gamma_{th}$. For example, when $\Psi_{fov} = 60°$ and $L = 3$ m, the VLC networks with $\gamma_{th} = 0$ dB and $-5$ dB fulfill the requirement on the outage probability if $\epsilon \geq 0.164$ and $\epsilon \geq 0.0096$, respectively.

In Fig. 2.6, we examine the feasible regions of $\xi$ and $\lambda_V$ when $\gamma_{th} = -5$ dB, $\epsilon = 0.25$, $L = 4$ m and $\Psi_{fov} = 60°$. In this case, as shown in Figs. 2.1 and 2.3, the VLC network outage probability is a decreasing function of $\lambda_V$, and VLC networks and RF networks can achieve the target outage probability $\epsilon = 0.25$. We observe that

**Table 2.2** RF network parameters [34]

| Symbol | $P_M$ | $P_M^C$ | $\Delta_M$ | $P_S$ | $P_S^C$ | $\Delta_S$ | $\alpha$ | $\sigma_R^2$ |
|--------|-------|---------|------------|-------|---------|------------|----------|--------------|
| Value  | 20 W  | 780 W   | 4.7        | 0.126 W | 13.6 W | 4          | 4        | $-100$ dBm   |

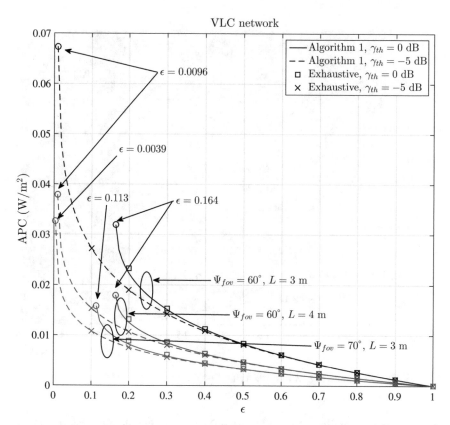

**Fig. 2.5** APC of the VLC networks as a function of $\epsilon$ [7]

the VLC networks and the RF networks respectively satisfy the target performance when $\lambda_V \geq 0.0092$ and $\xi \geq 3.2 \times 10^{-7}$. Since the outage probability of RF-optical HetNets is always lower than the outage probabilities of the standalone RF networks and the standalone VLC networks, as seen in (2.11), the target outage probability can be fulfilled in the RF-optical HetNets when $\lambda_V$ and $\xi$ are respectively smaller than 0.0092 and $3.2 \times 10^{-7}$. Therefore, as in Fig. 2.6, the feasible regions of $\xi$ and $\lambda_V$ for the RF-optical HetNets are wider than that of the RF networks and the VLC networks. In this context, by integrating the VLC networks with the RF networks, one can achieve a lower APC by judiciously selecting the values of $\lambda_V$ and $\xi$.

Figure 2.7 illustrates the feasible regions of $\xi$ and $\lambda_V$ when $\gamma_{th} = 0$ dB, $\epsilon = 0.25$, $L = 4$ m, and $\Psi_{fov} = 60°$. Under this setting, as observed in Figs. 2.1 and 2.3, RF networks cannot achieve the requirement on the outage probability and the VLC network outage probability is a convex function with respect to $\lambda_V$. Here, the two points $\lambda_V = 0.01$ and $0.028$ indicate the values of $\lambda_V$ that make the VLC network outage probability equal to $\epsilon = 0.25$. Thus, the feasible region of $\lambda_V$ for the VLC networks is $\lambda_V \in [0.01, 0.028]$ and $\lambda_V$ that minimizes the APC while achieving the

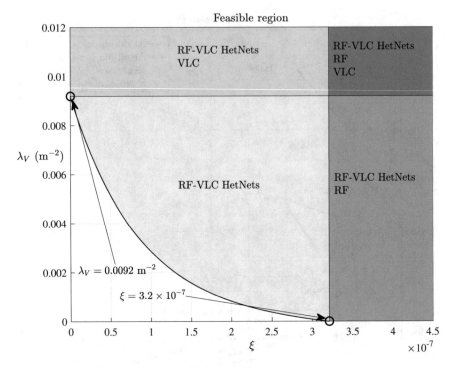

**Fig. 2.6** Feasible regions of $\xi$ and $\lambda_V$ when $\gamma_{\text{th}} = -5$ dB

target outage probability is given by $\lambda_V = 0.01$. Note that although the standalone RF networks cannot fulfill the target performance, by integrating the RF networks with the VLC networks, one can satisfy the target outage probability with values of $\lambda_V$ that are lower than 0.01, and this results in a reduced APC.

In Fig. 2.8, we present the APC of RF-optical HetNets with the introduced algorithms and the exhaustive search algorithms identifying the optimal densities $\lambda_M$, $\lambda_S$ and $\lambda_V$ by exhaustively comparing all combinations of the densities based on the tight approximation in (2.17). First, it is shown that the introduced algorithms experience negligible gaps compared to the exhaustive search algorithms and with much reduced complexities. As expected, the APC decays when the minimum performance requirement is relaxed, i.e., when $\epsilon$ increases or $\gamma_{\text{th}}$ decreases. Since RF signals cover wider ranges than VLC signals, when standalone RF networks fulfill the target performance, the required densities of RF BSs are much lower than that of optical BSs. Hence, in this case, for efficient power consumption, $\lambda_V$ becomes very small. However, small values of $\epsilon$, which are suitable for practical scenarios but cannot be satisfied by the standalone RF networks, are achievable when the VLC networks are integrated with the RF networks. In addition, RF-optical HetNets have lower APCs compared to the VLC networks, and the APC of the hybrid RF-optical networks is further reduced when the number of tiers of the RF BSs becomes

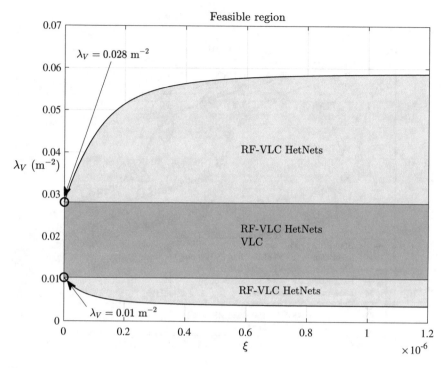

**Fig. 2.7** Feasible regions of $\xi$ and $\lambda_V$ when $\gamma_{th} = 0$ dB

larger. For example, when $\gamma_{th} = 0$ dB, the RF-optical HetNets fulfill the target performance $\epsilon \in [0.09,\ 0.19]$, which is not achievable in both standalone RF and VLC networks, while introducing even lower APCs than the standalone VLC networks when $\epsilon \geq 0.19$. From these observations, we can conclude that the RF-optical HetNets are more energy efficient than the standalone RF and VLC networks when the BS densities are properly determined.

## 2.6   Summary

This chapter investigated the deployment scenarios of standalone VLC networks and RF-optical HetNets where the RF networks are composed of two-tiers of BSs, i.e., MBSs and SBSs, and the MBSs, SBSs, and optical BSs follow independent PPPs. The goal of this chapter was to identify the optimal MBS, SBS, and optical BS densities that minimize the APC while satisfying a constraint on the SINR outage probability. The major findings of this chapter are summarized as follows:

- Since a practical PD's half of FOV is less than 90°, a user only receives the VLC signals transmitted by the optical BSs within a certain radius. Hence, the user may

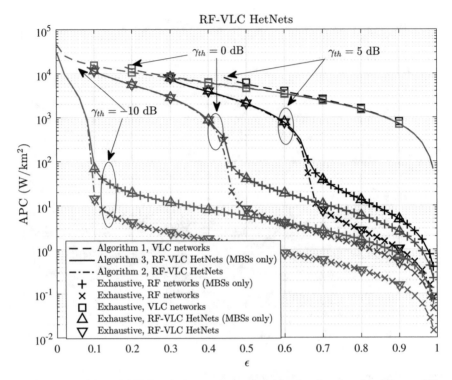

**Fig. 2.8** APC of the hybrid RF-optical networks as a function of $\epsilon$ when $L = 4$ and $\Psi_{\text{fov}} = 60°$

not be associated with an optical BS if the minimum distance between the user and its closest optical BS is larger than that radius. On the other hand, compared to the case when the half of FOV is 90°, the interference from nearby optical BSs becomes weaker when the half of FOV is smaller than 90°. In this case, the conventional method in [21], which computes the approximation of the PDF of the interference as a sum of Gamma distributions, cannot be directly employed since all moments of the interference are unknown. Therefore, in this chapter, by introducing a scaled density of the optical BSs that reflects the impact of the limited FOV on the interference, we managed to derive approximations of the outage probability of VLC networks which predict well the actual performance and present low complexities.

- From the observation that the outage probability of VLC networks is either a decreasing function or a convex function with respect to the density of optical BSs, we presented an algorithm to find the density of optical BSs that minimizes the APC of VLC networks while guaranteeing a minimum requirement on the outage probability. The introduced algorithm is based on our approximation of the outage probability and employs efficient one-dimensional line search methods, i.e., the golden section search method and the bisection method [44].

- Finally, we have examined the APC minimization problem for RF-optical HetNets with three optimization variables, namely, the densities of MBS, SBS, and optical BS. By exploiting the characteristics of the outage probabilities of RF networks and VLC networks, we reformulated the problem as a single-variable problem that is solved efficiently via line search algorithms. The simulation results corroborate the fact that the introduced algorithms for VLC networks and RF-optical HetNets achieve almost identical performance as the optimal benchmarks, however, with a much reduced complexity.
- RF-optical HetNets can satisfy target outage probabilities that cannot be satisfied by standalone RF and VLC networks. Furthermore, this can also be achieved with even lower APC than standalone networks if optimal densities of BSs from RF and optical networks are selected.

# References

1. P.H. Pathak et al., Visible light communication, networking, and sensing: a survey, potential and challenges. IEEE Commun. Surv. Tuts. **17**, 2047–2077 (2015)
2. H. Haas et al., What is LiFi? J. Lightw. Technol. **34**, 1533–1544 (2016)
3. L. Feng et al., Applying VLC in 5G networks: architectures and key technologies. IEEE Netw. **30**, 77–83 (2016)
4. M. Ayyash et al., Coexistence of WiFi and LiFi toward 5G: concepts, opportunities, and challenges. IEEE Commun. Mag. **54**, 64–71 (2016)
5. M. Ismail et al., *Green Heterogeneous Wireless Networks*, 1st edn. (Wiley, Hoboken, 2016)
6. M. Haenggi, *Stochastic Geometry for Wireless Networks* (Cambridge University Press, Cambridge, 2012)
7. J. Kong et al., Energy efficient optimization of base station intensities for hybrid RF/VLC networks. IEEE Trans. Wirel. Commun. **18**(8), 4171–4183 (2019)
8. J. Wang et al., Tight bounds on channel capacity for dimmable visible light communications. J. Lightw. Technol. **31**, 3771–3779 (2013)
9. R. Jiang et al., A tight upper bound on channel capacity for visible light communications. IEEE Commun. Lett. **20**, 97–100 (2016)
10. L. Wu et al., Adaptive modulation schemes for visible light communications. J. Lightw. Technol. **33**, 117–125 (2015)
11. L. Zeng et al., High data rate multiple input multiple output (MIMO) optical wireless communications using white LED lighting. IEEE J. Sel. Areas Commun. **27**, 1654–1662 (2009)
12. T.V. pham, h. le-minh, a.t. pham, multi-user visible light communication broadcast channels with zero-forcing precoding. IEEE Trans. Commun. **65**, 2509–2521 (2017)
13. H. Marshoud et al., On the performance of visible light communication systems with non-orthogonal multiple access. IEEE Trans. Wireless Commun. **16**, 6350–6364 (2017)
14. J. Hou, D.C. O'Brien, Vertical handover-decision-making algorithm using fuzzy logic for the integrated Radio-and-OW system. IEEE Trans. Wireless Commun. **5**, 176–185 (2006)
15. X. Li, R. Zhang, L. Hanzo, Cooperative load balancing in hybrid visible light communications and WiFi. IEEE Trans. Commun. **63**, 1319–1329 (2015)
16. Y. Wang et al., Optimization of load balancing in hybrid LiFi/RF networks. IEEE Trans. Commun. **65**, 1708–1720 (2017)
17. X. Wu, M. Safari, H. Haas, Access point selection for hybrid Li-Fi and Wi-Fi networks. IEEE Trans. Commun. **65**, 5375–5385 (2017)

18. M. Kashef et al., Energy efficient resource allocation for mixed RF/VLC heterogeneous wireless networks. IEEE J. Sel. Areas Commun. **34**, 883–893 (2016)
19. H. Zhang et al., Energy efficient subchannel and power allocation for software defined heterogeneous VLC and RF networks. IEEE J. Sel. Areas Commun. **36**, 658–670 (2018)
20. Cheng Chen, Dushyantha A. Basnayaka, Harald Haas, Downlink performance of optical attocell networks. J. Lightw. Technol. **34**, 137–156 (2016)
21. D.A. Basnayaka, H. Haas, Design and analysis of a hybrid radio frequency and visible light communication system. IEEE Trans. Commun. **65**, 4334–4347 (2017)
22. L. Yin, H. Haas, Coverage analysis of multiuser visible light communication networks. IEEE Trans. Wirel. Commun. **17**, 1630–1643 (2018)
23. H. Tabassum, E. Hossain, Coverage and rate analysis for coexisting RF/VLC downlink cellular networks. IEEE Trans. Wirel. Commun. **17**, 2588–2601 (2018)
24. A. Surampudi, R.K. Ganti, Interference characterization in downlink Li-Fi optical attocell networks. J. Lightw. Technol. **36**, 3211–3228 (2018)
25. Y.S. Eroğlu, Y. Yapici, and I. Güvenç. Impact of random receiver orientation on visible light communications channel. IEEE Trans. Commun. **67**, 1313–1325 (2019)
26. J. Gil-Pelaez, Note on the inversion theorem. Biometrika **38**(3/4), 481–482 (1951)
27. J.G. Andrews, F. Baccelli, R.K. Ganti, A tractable approach to coverage and rate in cellular networks. IEEE Trans. Commun. **59**, 3122–3134 (2011)
28. H.S. Dhillon et al., Modeling and analysis of -tier downlink heterogeneous cellular networks. IEEE J. Sel. Areas Commun. **30**, 550–560 (2012)
29. H-S. Jo et al., Heterogeneous cellular networks with flexible cell association: a comprehensive downlink SINR analysis. IEEE Trans. Wirel. Commun. **11**, 3484–3495 (2012)
30. D. Cao, S. Zhou, Z. Niu, Optimal combination of base station densities for energy-efficient two-tier heterogeneous cellular networks. IEEE Trans. Wirel. Commun. **12**, 4350–4362 (2013)
31. Y.S. Soh et al., Energy efficient heterogeneous cellular networks. IEEE J. Sel. Areas Commun. **31**, 840–850 (2013)
32. R. Cai, W. Zhang, P.C. Ching, Cost-efficient optimization of base station densities for multitier heterogeneous cellular networks. IEEE Trans. Wirel. Commun. **15**, 2381–2393 (2016)
33. J. Peng, P. Hong, K. Xue, Energy-aware cellular deployment strategy under coverage performance constraints. IEEE Trans. Wirel. Commun. **14**, 69–80 (2015)
34. E. Mugume, D.K.C. So, User association in energy-aware dense heterogeneous cellular networks. In: IEEE Trans. Wirel. Commun. **16**, 1713–1726 (2017)
35. Y. Zhuang et al., A survey of positioning systems using visible LED lights. IEEE Commun. Surv. Tuts. **20**, 1963–1988 (2018)
36. A.R. Ndjiongue, H.C. Ferreira, An overview of outdoor visible light communications. Trans. Emerg. Telecommun. Technol. **29**, 1–15 (2018)
37. B.G. Guzmán et al., Downlink performance of optical OFDM in outdoor visible light communication. IEEE Access **6**, 76854–76866 (2018)
38. I.E. Lee, M.L. Sim, F.W.L. Kung, Performance enhancement of outdoor visible-light communication system using selective combining receiver. IET Optoelectron. **3**, 30–39 (2009)
39. N. Lourenço et al., Visible light communication system for outdoor applications, in *Proceedings of IEEE CSNDSP*
40. J.G. Andrews et al., What will 5G be? IEEE J. Sel. Areas Commun. **32**, 1065–1082 (2014)
41. C-H. Lee, C-Y. Shih, Coverage analysis of cognitive femtocell networks. IEEE Wirel. Commun. Lett. **3**, 177–180 (2014)

42. Z. Yazdanshenasan et al., Poisson hole process: theory and applications to wireless networks. IEEE Trans. Wirel. Commun. **15**, 7531–7546 (2016)
43. I.S. Gradshteyn, I.M. Ryzhik, *Table of Integrals, Series and Products*, 7th edn. (Academic Press, 2007)
44. E.K.P. Chong, S.H. Zak, *Introduction to Optimization*, 4th edn. (Wiley, Hoboken, 2013)

# Chapter 3
# Realization and Dataset Generation for Mobile Indoor Channels

**Abstract** High frequency bands in 5G and beyond (5G+) heterogeneous networks (HetNets) are challenged by link instabilities induced by users' mobility, which are attributed to the wave propagation characteristics in such bands. Hence, good understanding of channel characteristics in dynamic high frequency bands plays a vital role in developing robust resource management strategies in 5G+ HetNets. Unfortunately, it is quite difficult to collect accurate indoor channel data in dynamic environments. In addition, indoor trajectory datasets are not publicly available. In order to address such limitations, this chapter introduces an analytical framework to model realistic human mobility patterns in dynamic indoor environments. The presented model abstracts the nature of human behavior by integrating both macro and micro mobility patterns. These patterns are then used to realize the spatio-temporal characteristics of wireless channels under long-term environment-confined mobility. We apply this framework to characterize the wireless indoor channel for optical wireless networks whose downlink adopts visible light and uplink adopts infrared light. Our results demonstrate that the indoor layout and the user's environment-confined mobility pattern significantly impact the line-of-sight (LOS) dynamics but present limited impact on non-LOS components. The presented framework can be used to generate realistic wireless channel datasets that could be further employed to develop data-driven resource management strategies in 5G+ HetNets.

## 3.1 Introduction

Several proposals have been made to implement and commercialize visible light communication (VLC) networks, which are also referred to as Light-Fidelity (Li-Fi) networks [1]. Due to constraints related to the user equipment (UE) orientation and power, such networks can support VLC in the downlink and infrared (IR) communications in the uplink. In order to carry out proper management of resources in such networks, good understanding of the channel characteristics is required. Due to the propagation nature of light, the optical channel is susceptible to burst occlusion and displacement [2]. Therefore, the details of the environment and the UE trajectory

Z.-Y. Wu et al., *Efficient Integration of 5G and Beyond Heterogeneous Networks*,
https://doi.org/10.1007/978-981-15-6938-8_3

and orientation dominate the variations in the channel states. It should be highlighted that the wavelength of light is only a few hundred nanometers, and thus, it is almost impossible for light to bypass the macro obstacles by diffraction. This results in frequent outages in line-of-sight (LOS) transmissions when the user is moving in a complex indoor environment. Hence, the specific motion trajectory determines the quality of the LOS transmission. In other words, the observed pattern of the LOS channel is a projection of the environment-confined mobility pattern.

Unfortunately, our mastery of indoor mobility patterns is very limited. One of the reasons is that the source of mobility, i.e., accurate indoor motion trajectory data, is not readily available [3]. Another factor is the lack of high-quality channel datasets. From the perspective of practical measurements, if participants use smart terminals to collect a sizable dataset, it is hard to achieve sufficient accuracy of optical intensity measures in order to capture meaningful statistics under mobility. In this chapter, we aim to establish a framework that is capable of producing trajectories in indoor scenarios, by means of which mobile optical channels can be properly characterized. The presented mobility model can also be adopted to generate channel datasets for high frequency bands such as mmWave and Terahertz, once the appropriate channel model is applied to the developed indoor trajectories [4].

### 3.1.1 Background

Mobility support in wireless channels involves understanding the patterns of the UE's mobility as well as the wave propagation features. In the following, we focus on optical channels as one example of high frequency bands to be adopted in 5G and beyond (5G+) heterogeneous networks (HetNets). Hence, we discuss optical channel models along with mobility models.

#### 3.1.1.1 Channel Models

The first step in modeling an optical channel is to study the radiation pattern of a light source. Moreno et al. [5] introduced a general and accurate representation for the radiation pattern of a light-emitting-diode (LED). The pattern was expressed as the sum of two or three Gaussian or cosine-power functions. However, in a lighting-supported environment, LEDs are usually grouped as arrays rather than as single point sources. Ding et al. [6] compared the channel characteristics of both the simplified point-source model and several practical cases using various scales of arrays. As for the generation of the channel impulse response (CIR), Kahn et al. [7] summarized the modeling and simulation methods of ray tracing for wireless infrared communications. Schulze in [8] presented a frequency-domain simulation method for an infinite number of reflections. More detailed channel models for VLC were given by Lee et al. [9] and were further developed by Jungnickel et al. [10]. The ray tracing methods for efficient non-line-of-sight (NLOS) channel computation were introduced by Chen et al. [11]. The aforementioned efforts provide a solid foundation for static channel models and analysis. On the other hand, limited research works considered mobile scenarios. For instance, the mobile channel model introduced by Miramirkhani et al. [12] exhibits significant variations in the received power along a straight line trajectory. Also, Chvojka et al. [13] presented analytical and experimental results to model the impact of random movements in various indoor scenarios, where both shadowing and blocking effects were investigated. More recently, Soltani et al. [14] introduced an orientation-based random waypoint (ORWP) mobility model. However, existing channel models in a dynamic environment usually assume over-simplified mobility models which cannot reproduce a realistic and comprehensive human trajectory (e.g., [12–14]).

### 3.1.1.2 Mobility Models

Although random walk models such as random waypoint used in e.g. [14, 15] are popular and straightforward, they fail to reflect human mobility accurately from the perspective of the nature of human movement instincts and the features of dynamic UE orientation. On the other hand, bounded Lévy-walk with heavy-tail distance and pause-time distributions is statistically more suitable to model human mobility [16]. Different from the behaviors in conventional random walk models, real human trajectories exhibit strong tendency to return to the visited locations and manifest through recurrence and temporal periodicity in human mobility [17]. From this perspective, the human trajectory is predictable. Besides, more importantly, the vital statistical characteristics of individual trajectories are largely indistinguishable regardless of their spatio-temporal scales [17]. Statistical results collectively confirm that human mobility is characterized by scale-freedom in which, however, scales, human trajectories exhibit similar patterns to those of a Lévy-walk [16]. Therefore, the same statistical characteristics of Lévy-walk and return tendency can be both deployed in indoor environments after reducing the scale.

Unfortunately, existing mobility models do not capture macro and micro patterns that best reflect human behavior in an indoor scenario. In this context, while Lévy-walk and return tendency-based models can be used to describe the next destination point for a mobile user (i.e., macro pattern), the detailed trajectory between the origin and destination (i.e., micro patterns) that accounts for path selection [18], steering forces [19], and orientation of UEs [14] should be captured.

Hence, in order to develop realistic wireless channel datasets for indoor dynamic environments, we first present in this chapter an analytical framework to develop trajectories that best reflect human mobility patterns within indoor settings. Proper channel models will then be applied on the developed trajectories to generate optical channel datasets in dynamic indoor environments. The generated dataset will be used in the next chapters for developing data-driven resource management strategies in 5G+ HetNets.

### *3.1.2  Chapter Organization*

The rest of this chapter is organized as follows. Section 3.2 presents a realistic indoor mobility model that captures both macro and micro human mobility patterns. Section 3.3 discusses a framework to generate the CIR data in a dynamic environment using the mobility model. Section 3.4 interprets how to collect key mobility parameters from the real-world and yields the system parameter setup to generate the channel dataset. Section 3.5 presents a set of results to characterize the optical wireless channel for both LOS and NLOS links. Conclusions are drawn in Sect. 3.5.

## 3.2  Mobility Model

Figure 3.1 illustrates an instance of a practical trajectory driven by the mobility of a human. The trajectory data is then employed to the synthesis of channel dataset. To generate such trajectories as shown in Fig. 3.1, we start by introducing a realistic mobility model that captures the human behavior at different spatio-temporal scales, namely:

- *Macro patterns*: These define the next destination points and are modeled as a semi-Markov renewal process subject to bounded Lévy-walk and return regularity.
- *Micro patterns*: These define the detailed trajectory between the origin and destination points along with the UE orientation and are represented by a large-scale trajectory subject to the shortest path, a small-scale trajectory featuring steering behavior, and a random terminal orientation.

The details of these patterns are discussed next.

### *3.2.1  Macro Patterns*

The macro-scale mobility patterns are described using a semi-Markov renewal process [20]. Through statistical analysis [20], the return regularity and truncated Lévy-walk dominate the transition probabilities among different destination points (states) in this process.

The destination set is defined consistently with various types of furniture and entrances within a room. Each type of furniture has a particular mean sojourn time, which defines the time a user dwells in the range of the furniture. At this level, it is assumed that the semi-Markov renewal process is time-homogeneous and represented by $\{(X_n, T_n) : n \geq 0\}$ with discrete state space of residential destinations $\mathcal{L} = \{1, 2, 3..., L\}$ (the furniture and entrances) where $T_n$ is the time of the $n$th transition, $X_n$ is the state at the $n$th transition [20], and $L$ stands for the number of

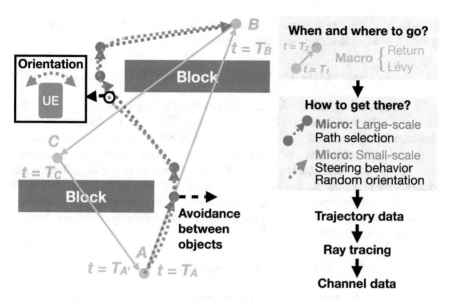

**Fig. 3.1** Diagram of a realistic human mobility, which can be utilized for generating mobile channel dataset

the destinations including all the resident nodes and entrances. The state of a semi-Markov process represents each destination, and traveling from one destination to another is a state transition.

The choice of each destination is first determined by return regularity, by which the user decides to return to a particular destination. It is important to note that return regularity is different from the general random walk model. In contrast with the smooth asymptotic behavior predicted for random walks [17], in realistic human trajectories, there always exists a recurrence and temporal periodicity inherent to human mobility as returning to a specific place with a strong tendency to locations they visited before in a particular time. For the sake of characterizing this return tendency, the semi-Markov process should cover the stage-of-trajectory $T_d$ (e.g., morning, noon, afternoon, and night) during each transition [21]. Accordingly, we consider the time-homogeneous semi-Markov kernel, which denotes the probability of transition into state $j \in \mathcal{L}$ within $t$ units of time immediately after the transition into state $i \in \mathcal{L}$ as

$$\zeta_{i,j}^{(T_d)}(t) = p_{i,j}^{(T_d)} \mu_{i,j}^{(T_d)}(t) = \Pr\{X_{n+1} = j, T_{n+1} - T_n \le t | X_n = i, T_{n+1}, T_n \in T_d\},$$
$$(3.1)$$

where $T_n$ and $T_{n+1}$ are the occurrence time of state $X_n$ and $X_{n+1}$, $p_{i,j}^{(T_d)}$ is the transition probability from start location $i$ to destination $j$ defined as

$$p_{i,j}^{(T_d)} = \lim_{t \to \|T_d\|_\infty} \zeta_{i,j}^{(T_d)}(t) = \Pr\{X_{n+1} = j | X_n = i, T_{n+1}, T_n \in T_d\} \qquad (3.2)$$

In (3.1), $\mu_{i,j}^{(T_d)}$ denotes the time distribution including: a) the duration of dwelling at state $i$, $t_s$ and b) the transition into the destination $j$, $t_{i \to j}$. We have

$$\mu_{i,j}^{(T_d)}(t) = \Pr\{T_{n+1} - T_n \le t | X_{n+1} = j, X_n = i, T_{n+1}, T_n \in T_d\} \quad (3.3)$$

In the general movement during the absence of the return action, the bounded Lévy-walk plays the decisive role in each transition. The truncated Pareto distribution characterizes this pattern. Statistical results collectively confirm the scale-freedom characterization of human mobility [16, 17]. Human movements share similar patterns on any scale since the key statistical characteristics of re-scaled individual trajectories are identical. Hence, even for indoor trajectories, this truncated heavy-tail distribution contributes prominently. Lévy-walk is a kind of continuous-time random walk. We consider a random walker and choose the joint space-time probability density function (PDF):

$$\Phi\left(r_{i \to j}, t_{i \to j}\right) = \phi\left(t_{i \to j} | r_{i \to j}\right) p(r_{i \to j}), \quad (3.4)$$

where $p(r_{i \to j})$ is the probability that a step of length $r_{i \to j}$ occurs and $\phi(t_{i \to j} | r_{i \to j})$ is the conditional probability density that such step takes $t_{i \to j}$ time in movement. When $p(r_{i \to j})$ is a heavy-tailed distribution such as Pareto distribution, the mobility specified by $\phi(t_{i \to j} | r_{i \to j})$ is a Lévy-walk. Unifying this definition with our previous discussion on semi-Markov process, we assign the destination $j$ with respect to starting point $i$ within Euclidean distance $r_{i \to j}$ via the PDF of a truncated Pareto distribution as

$$p_{i,j} = p(r_{i \to j}) = \frac{\alpha(r_{\min})^\alpha r_{i \to j}^{-(\alpha+1)}}{1 - (r_{\min}/r_{\max})^\alpha}, \quad (3.5)$$

in which $\alpha$ is a positive parameter, $r_{\min} = \min_{\forall a,b \in \mathcal{L}} \{r_{a \to b}\}$ and $r_{\max} = \max_{\forall a,b \in \mathcal{L}} \{r_{a \to b}\}$. Also, the sojourn time $t_s$ that a user dwells in a residential location $i$ is also subjected to a truncated Pareto distribution [16] as

$$p_{t_s}(t_s; \beta, t_{s,\max}^{(i)}, t_{s,\min}^{(i)}) = \frac{\beta\left(t_{s,\min}^{(i)}\right)^\beta t_s^{-(\beta+1)}}{1 - \left(t_{s,\min}^{(i)}/t_{s,\max}^{(i)}\right)^\beta}, \quad (3.6)$$

where $\beta$ corresponds to a positive parameter, $t_{s,\max}$ and $t_{s,\min}$ are the maximum and minimum sojourn times at the current location, respectively.

However, the corresponding time steps $t_{i \to j}$ in $\phi\left(t_{i \to j} | r_{i \to j}\right)$ are determined by a two-scale trajectory model regarding the transition $i \to j$, which will be discussed in the following subsection. All the statistical results we are referring to are based on data collection using the distance between any two resident spots rather than the actual trajectory length in detail. Therefore, to stick with the status quo, the step distance in (3.5) is not the exact length in a real trace, but the straight line distance.

## 3.2.2 Micro Patterns

The micro pattern details the trajectories between every starting point to its destination as $\Theta(i, j, t)$, based on a two-level model: large-scale and small-scale descriptions. The large-scale model describes the movement path where the user follows a sequence of intermediate target nodes, while the small-scale model takes into account the 3-dimensional (3D) successive user positions and UE orientations.

In this chapter, we only focus on mobile states, where the link quality is under serious challenge due to mobility. As aforementioned, the action of choosing a destination is dominated by the bounded Lévy-walk process and return tendency in the macro pattern, so in this subsection, large-scale mobility is defined by the shortest path according to the Dijkstra algorithm [18] to reach the destination. The graph of resident nodes (destinations) $\mathcal{G}_r$ for furniture and path nodes (grid) $\mathcal{G}_p$ for the endpoints of segments in the path are established. Graph $\mathcal{G}_p$ represents all the possible transitions between furniture and is arranged for the clearance of the impassable areas. Meanwhile, $\mathcal{G}_p$ has all the resident nodes for covering the starts and ends of every trace. Extra nodes are associated with the doors for engaging the entering and exiting mobility as $\mathcal{G}_r \subset \mathcal{G}_p$. These graphs ought to keep the user from an unrealistic trajectory. Graph $\mathcal{G}_r$ allows the user to choose an item of furniture as a destination following the macro pattern, and $\mathcal{G}_p$ ensures a path to each destination. A feasible path $\mathcal{V}_{i \to j}$ from a resident location $i$ to a location $j$ is obtained using the Dijkstra shortest path algorithm $\mathcal{D}$:

$$\mathcal{V}_{i \to j} = \mathcal{D}(i, j), \tag{3.7}$$

which is a sequence of successive path nodes.

In addition to the given path $\mathcal{V}_{i \to j}$, when encountering the departure, human mobility in a smaller scale that encodes the detailed trajectories is described by the steering behavior model presented in [19]. Steering behavior $\mathcal{S}$ is brought to bear on the interactions between the user and environment $\Omega$ so that one can reproduce a simple physical engine upon the path node sequence as

$$\Theta(i, j, t) = \mathcal{S}(\mathcal{V}_{i \to j}, \Omega). \tag{3.8}$$

Typically, the user is treated as a point particle that carries a point mass $\mathrm{m}$, maximum acceleration $a_{\max}$ and a maximum velocity of $v_{\max}$. Then, the very basic Newton's laws of motion can be applied, where the user is driven by several steering forces $\mathbf{F}(t)$ applied on their center of mass. At time step $t$, the acceleration $a(t)$ is renewed according to the resulting steering force applied on the user. The applied acceleration vector $\mathbf{a}(t)$ is given by $\mathbf{a}(t) = \mathbf{F}(t)/\mathrm{m}$. The resulting velocity $\mathbf{v}(t)$ is approximated by the Euler integration as an augment of the product of the current acceleration vector with the time interval $\delta_\tau$ to the previous velocity, i.e.,

$$\mathbf{v}(t) = \mathbf{v}(t - \delta_\tau) + \mathbf{a}(t)\delta_\tau. \tag{3.9}$$

The user's position $\mathbf{p}(t)$ is obtained by the Euler integration as an augment of the product of the current applied velocity vector with the time interval $\delta_\tau$ upon the previous position, i.e.,

$$\mathbf{p}(t) = \mathbf{p}(t - \delta_\tau) + \mathbf{v}(t)\delta_\tau. \tag{3.10}$$

On this basis, it is necessary to formulate the details of the steering forces to illustrate how the human-like behaviors might unfold. In this chapter, for the sake of simplicity, a single user is assumed, and thus only seek and avoidance behaviors are considered.

The seek behavior produces a seek force for attracting the user to each target node in the successive node sequence along the selected path. The seek force orients towards the distance vector $\mathbf{d}(t)$ between the intermediate target $\xi(t) \in \mathcal{V}_p$ and the actual position of the user $\mathbf{p}(t)$ as $\mathbf{d}(t) = \xi(t) - \mathbf{p}(t)$. Hence, the corresponding desired velocity vector $\mathbf{v}^d(t)$ is given by

$$\mathbf{v}^d(t) = \frac{\mathbf{d}(t)}{\|\mathbf{d}(t)\|} \frac{v_{\max}}{\delta_\tau} \tag{3.11}$$

where $\|\cdot\|$ represents the 2-norm of vector. Following the velocity difference between the actual velocity and the desired velocity, the user is driven by the seeking force $\mathbf{F}_s(t)$ as

$$\mathbf{F}_s(t) = \mathrm{m}\frac{\mathbf{v}^d(t) - \mathbf{v}(t)}{\delta_\tau}. \tag{3.12}$$

Contrary to seeking, while approaching the destination, the user also slows down to end the period of mobility. This behavior is modeled by arrival force $\mathbf{F}_a(t)$ that is a reverse to $\mathbf{F}_s(t)$ but with a threshold radius to ignore long range effects.

On the other hand, to repulse a user from penetrating insurmountable areas such as the layout of furnishings, an avoidance force ought to be applied upon the user. Considering the perpendicular distance from the present position to one of the surfaces of obstacles, the avoidance force $\mathbf{F}_o(t)$ is given by

$$\mathbf{F}_o(t) = \mathrm{m}\frac{\hat{\mathbf{n}}}{d_w(t)} \frac{\|\mathbf{v}(t) - \mathbf{v}(t - \delta_\tau)\|}{\delta_\tau}, \tag{3.13}$$

in which $d_w$ is the perpendicular distance to blockage surface $w$, and $\hat{\mathbf{n}}$ denotes an orthogonal unit vector against $w$.

By taking all behaviors into account, we express the resulting force as

$$\mathbf{F}(t) = \mathbf{F}_s(t) + \gamma_o \sum_w^{\mathcal{W}} \mathbf{F}_o(t) + \gamma_a \mathbf{F}_a(t), \tag{3.14}$$

where $\gamma_o$ is the ratio of avoidance force to seek force, $\gamma_a$ stands for the ratio of arrival force to seek force, and $\mathcal{W}$ indicates the number of surfaces of all blockages with an effective distance threshold for the sake of neglecting the remote surfaces so as to improve the computational efficiency. Finally, we will have to limit the scale of the force:

$$\mathbf{F}(t) := \max\{\|\mathbf{F}(t)\|, \mathbb{m}a_{\max}\} \frac{\mathbf{F}(t)}{\|\mathbf{F}(t)\|}. \qquad (3.15)$$

The micro pattern of mobility also models random orientations of the UE [14]. The indoor optical wireless channel can be treated as a slowly-varying channel since its delay spread is typically on the order of nanoseconds, and the coherence time of the random orientation is in the order of hundreds of milliseconds [14, 22]. Rotation geometry for UE usually requires three elemental angles called yaw, pitch, and roll. The concatenated rotation matrix with respect to the Earth coordinate system indicates the rotated normalized vector represented in the spherical coordinate system with the polar angle $\vartheta$ and azimuth angle $\omega$. The polar angle is between the rotated normalized vector and the vector perpendicular to the ground plane; the azimuth angle stands for the projection of the rotated unit vector in the ground plane relative to the positive direction of meridian. The statistics from real-world [14, 22] show that the polar angle should be modeled as a Laplace distribution $\vartheta \sim \text{Laplace}(\mu_s, b_s)$ for the sitting activity and a Gaussian distribution $\vartheta \sim \mathcal{N}(\mu_w, \sigma_w)$ for the walking activity. In this chapter, the azimuth angle is obtained directly from user trajectories. Meanwhile, key parameters $\mu_s$, $b_s$, $\mu_w$, and $\sigma_w$ are resolved from measurements in real-world according to the methodology in [14].

## 3.3   Generation of Mobile Channel Data

Figure 3.2 shows the hierarchy of the introduced framework in this chapter to generate the channel dataset subject to the realistic human mobility.

Following this hierarchy, we first generate the complete trajectories and mobility data $\mathbf{M}(t)$ following the model described in Sect. 3.2 that integrates the macro and micro patterns. The procedures of producing macro mobility are given in Algorithm 3, where we organize the mobility into five statuses as follows: (1) "Return": the movements due to the return regularity; (2) "Arrive": the behaviors of choosing the sojourn duration and the next destination; (3) "Sojourn": the implementation of the sojourn behavior at resident nodes; (4) "Depart": the behaviors of choosing the trajectories $\Theta(i, j, t)$; (5) "Transit": the implementation of the chosen trajectories. Also, Algorithm 4 offers the details on producing the micro mobility in the trajectories of $\Theta(i, j, t)$.

According to the framework of CIR generation illustrated in Fig. 3.2, we will use the data of UE position and orientation to synthesize the channel data in this chapter. We consider the blockage of opaque objects from furniture and user body before calculating the specific channel gain of each link. Standard procedures for judging the viability of each ray in advance are demonstrated as follow. In the aforementioned coordinate system, the receiver incidence angle $\theta$ is decided by $\vartheta$ and $\omega$ and the rel-

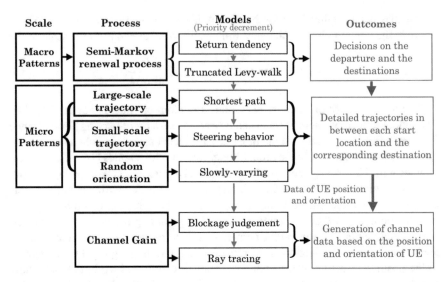

**Fig. 3.2** Hierarchy of the introduced framework to generate CIR in dynamic optical wireless channels

ative angle between a transmitter and a receiver according to the configuration of the placement of the transceiver on the UE. Also, because we need to reduce the potential harm of the irradiation to the eyes from uplink and the probability of the blockages from the user body that will induce considerable outages, the transceiver on an UE should point forward the trace direction rather than towards the ceiling perpendicular to the surface of screen during mobility. Therefore, we define a direction vector for a UE's transceiver as $\mathbf{u}_\theta^{UL}(t) = (\sin \vartheta(t) \cos \omega(t), \sin \vartheta(t) \sin \omega(t), \cos \vartheta(t))$, and for a base station (BS) transceiver as $\mathbf{u}_\theta^{DL} = (0, 0, -1)$. Then, we get $\cos \theta(t) = \mathbf{u}_\theta^{UL}(t) \cdot \mathbf{u}_\theta^{DL} / (\|\mathbf{u}_\theta^{UL}(t)\| \cdot \|\mathbf{u}_\theta^{DL}\|)$.

Additionally, we express the transmission distance of LOS as $d_0(t) = \|\mathbf{p}(t) - \mathbf{p}_{BS(v)}\|$, where $\mathbf{p}_{BS(v)}$ corresponds to the location of BS $v \in \{1, \ldots, v, \ldots, V\}$ in which $V$ is the number of BSs. Hence, before computing the channel gain for a single ray, if any intersection between this ray with any of the surfaces including the furniture and the user body is found, we judge this ray as blocked. In this brief, we assume the user body and opaque objects are in the form of cuboids, and the blockage of a ray is detected as illustrated in Fig. 3.3. For each object and each ray, if there exists at least one view (vertical, side, front) from which the projection of the ray upon this view does not cross any edge of the projection of the surface upon this view, nor is encircled by the edges (i.e., neither the start and end points of the ray are located inside the area), then this ray passes this object.

Given a cubic environment $\Omega$ that occupies a certain volume, the overall time-varying impulse response $h_\Omega$ is defined as

---

**Algorithm** 3. State Machine for Generating Mobility

---

1: Initialize the room layout and the resident and path nodes
2: Initialize *EventList* for recording the statuses
3: $\mathbf{p}(t = 0) = \mathbf{p}_{door}$ as the user enters the door
4: *EventList*$(t = \delta_\tau) \leftarrow$ "*Sojourn*",
5: *EventList*$(t = 2\delta_\tau) \leftarrow$ "*Depart*", $t \leftarrow 2\delta_\tau$
6: Set the first destination $j$ according to $p(r_{i \rightarrow j} | \mathbf{p}_{door})$
7: **while** $\|\mathbf{p}(t) - \mathbf{p}_{door}\| > \epsilon$ **do**
8:     *EventList*$(t_{\text{return}}) \leftarrow$ "*Return*" according to (3.1)
9:     **switch** *EventList*$(t)$
10:       **case** "*Depart*" **do**
11:         $i \leftarrow \mathbf{p}(t)$
12:         Execute micro trace generation and get $t_{\text{arrival}}$, *EventList*, and $\Theta(i, j, t)$
13:         Update mobility status $\mathbf{M}(t) \leftarrow \Theta(i, j, t)$
14:       **case** "*Arrive*" **do**
15:         Update mobility status $\mathbf{M}(t) \leftarrow \Theta(i, j, t)$
16:         Generate $t_s \sim$ Pareto$(\beta, t_{s,\max}, t_{s,\min})$
17:         $t_{\text{departure}} \leftarrow t_{\text{arrival}} + t_s$
18:         Choose destination $j$ according to $p(r_{i \rightarrow j} | i)$  as (3.5)
19:         *EventList*$(t \sim t_{\text{departure}} - \delta_\tau) \leftarrow$ "*Sojourn*"
20:         *EventList*$(t_{\text{departure}}) \leftarrow$ "*Depart*"
21:       **case** "*Sojourn*" **do**
22:         Update sitting motion status $\mathbf{M}(t)$
23:       **case** "*Return*" **do**
24:         Reset *EventList* from time slot $t$
25:         $i \leftarrow \mathbf{p}(t), j \leftarrow \mathbf{p}_{\text{Return}}$
26:         Execute micro trace generation and get $t_{\text{arrival}}$, *EventList*, and $\Theta(i, j, t)$
27:         Update mobility status $\mathbf{M}(t) \leftarrow \Theta(i, j, t)$
28:       **case** "*Transit*" **do**
29:         Update mobility status $\mathbf{M}(t) \leftarrow \Theta(i, j, t)$
30:     **end switch**
31:     $t \leftarrow t + \delta_\tau$
32: **end while**
33: **return M**

---

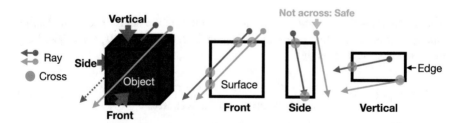

**Fig. 3.3** If a ray can safely pass the object, there must exist at least one view (vertical, side, front) from which the projection of the ray upon the surface of this view does not intersect with any edge. So in this example, the green ray does not have any intersection with the projection on the surface in the side view, unlike the red ray that intersects with edges from all views [23]

---

**Algorithm** 4. Generation of Micro Trajectory: $\Theta(i, j, t)$

---

1: $\mathbf{p}(t) \leftarrow i, t_i = t$
2: Get adjacent matrix for $i$ , $j$ from $\mathcal{G}_p$
3: Deploy Dijkstra algorithm: $\mathcal{V}_{i \rightarrow j} = \mathcal{D}(i, j)$
4: Set a margin $\epsilon$ for checking if the user arrives the range of a node
5: **while** $\|\xi(t) - j\| > \epsilon$ **do**
6:   **if** $\xi(t) = j$ **then**
7:     **if** $\|\mathbf{p}(t) - \xi(t)\| < \epsilon$ **then**
8:       **goto** *ending*
9:     **else**
10:       $\mathbf{F} \leftarrow \mathbf{F}_s(t) + \gamma_o \sum_w^W \mathbf{F}_o(t) + \gamma_a \mathbf{F}_a(t)$
11:     **end if**
12:   **else**
13:     **if** $\|\mathbf{p}(t) - \xi(t)\| < \epsilon$ **then**
14:       $\xi(t) \leftarrow \xi(t + \delta_\tau) \in \mathcal{V}_{i \rightarrow j}$
15:     **else**
16:       $\mathbf{F}(t) \leftarrow \mathbf{F}_s(t) + \gamma_o \sum_w^W \mathbf{F}_o(t)$
17:     **end if**
18:   **end if**
19:   Generate $\vartheta(t) \sim \mathcal{N}(\mu_w, \sigma_w)$
20:   Get $\mathbf{v}(t)$ and $\mathbf{p}(t)$
21:   $\Theta(i, j, t) \leftarrow (\mathbf{v}(t)|\mathbf{p}(t)|\vartheta(t))$
22:   $t \leftarrow t + \delta_\tau$
23: **end while**
24: *ending*:
25:   $t_{\text{arrival}} \leftarrow t$
26:   $EventList(t_i \text{ to } t_{\text{arrival}} - \delta_\tau) \leftarrow$ "Transit"
27:   $EventList(t_{\text{arrival}}) \leftarrow$ "Arrive"
28:   **return** $\Theta(i, j, t)$ and $EventList$

---

$$h_\Omega(t, \tau) = \sum_{k=0}^{\infty} h^{(k)}(t, \tau), \qquad (3.16)$$

where $k$ denotes the reflection index and $\tau$ represents the delay. The total channel gain is defined as the direct-current (DC) gain $H(t, 0) = \int_\infty h_\Omega(t, \tau)d\tau$. The LOS impulse response is given by

$$h^{(0)}(t, \tau) = \begin{cases} \frac{A_R}{d_0^2(t)} \frac{(m+1)}{2\pi} \cos^m \psi(t) \cos\theta(t) T_S(\theta(t)) \delta\left(\tau - \frac{d_0(t)}{c}\right), & \text{if } 0 \leq \theta(t) \leq \Psi \\ 0, & \text{if } \theta(t) > \Psi \text{ or ray is blocked}, \end{cases} \qquad (3.17)$$

where $A_R$ is the sensor area, $\psi$ stands for the angle of irradiance, $\Psi$ is the receiver's field-of-view (FOV), $c$ means the speed of light, and $T_S(\theta)$ denotes the overall transmission response of the optical system and is treated as 1 in this chapter. Besides, the mode number $m$ is related to half power angle $\Phi_{1/2}$ via $m = -\ln 2 / \ln \cos \Phi_{1/2}$. For the diffused NLOS links, a recursive ray-tracing method [24] is adopted and microelements $\{\varepsilon\}$ representing differential of reflective areas are considered. For path $k > 1$, the intermediate propagation from a small cell $\varepsilon_n$ to another one $\varepsilon_m$ contributes to the NLOS response recursively as

$$h^{(k)}(t, \tau) = \sum_{m=1}^{M} \sum_{n=1}^{N} h_{\varepsilon_m \to \varepsilon_n}^{(k-1)}(t, \tau) * \rho_\varepsilon h_{\varepsilon_n \to R}^{(0)}(t, \tau), \tag{3.18}$$

in which $R$ denotes the receiver and $\rho_\varepsilon$ is the reflectivity of the surface to which the element belongs. In addition, we divide the reflective surfaces into $N$ elements with the FOV as $90°$ [9, 13]; especially for $k = 1$ from the transmitter $S$, we have propagation $S \to \varepsilon$ instead of the ones among other elements expressed as

$$h^{(1)}(t, \tau) = \sum_{n=1}^{N} h_{S \to \varepsilon_n}^{(0)}(t, \tau) * \rho_\varepsilon h_{\varepsilon_n \to R}^{(0)}(t, \tau), \tag{3.19}$$

where

$$h_{S \to \varepsilon}^{(0)}(t, \tau) = \begin{cases} \frac{A_\varepsilon}{d_{S,\varepsilon}^2(t)} \frac{(m+1)}{2\pi} \cos^m \psi_{S,\varepsilon}(t) \cos \theta_{S,\varepsilon}(t) \delta \left( \tau - \frac{d_{S,\varepsilon}(t)}{c} \right), \\ \qquad\qquad\qquad \text{if } 0 \le \theta_{S,\varepsilon}(t) \le 90° \\ 0, \text{ if } \theta_{S,\varepsilon}(t) \ge 90° \text{ or blocked.} \end{cases} \tag{3.20}$$

$$h_{\varepsilon \to R}^{(0)}(t, \tau) = \begin{cases} \frac{A_R}{d_{\varepsilon,R}^2(t)} \frac{(m+1)}{2\pi} \cos^m \psi_{\varepsilon,R}(t) \cos \theta_{\varepsilon,R}(t) T_S \left( \theta_{\varepsilon,R}(t) \right), \\ \qquad\qquad\qquad \text{if } 0 \le \theta_{\varepsilon,R}(t) \le 90° \\ 0, \text{ if } \theta_{\varepsilon,R}(t) \ge 90° \text{ or blocked.} \end{cases} \tag{3.21}$$

The computational cost of this recursive approach is $N^K$, where $N$ is the number of elements into which the reflecting surfaces are divided. Meanwhile, the Monte-Carlo ray-tracing algorithm can be more efficient as the computational cost can be $K \cdot N_{Ray} \cdot N_F$ [25], where $N_{Ray}$ stands for the number of rays that is set for accuracy, and $N_F$ corresponds to the number of triangles for defining the geometry. However, in this chapter, we rely on a huge dataset to capture the spatio-temporal evolution of the channel statistics at a macro level; therefore, we consider $K = 1$ as the contribution of the higher order reflections ($K \ge 2$) is very small compared to that of the LOS and $K = 1$ components as demonstrated in [26]. As such, the computational cost difference between these two methods is very small, and eventually we choose the deterministic one with more accurate outcomes.

The model we presented in this chapter is applicable to both VLC and IR. However, in IR communications, the reflectance of materials is typically modeled as a constant. On the other hand, the reflectance of materials in the VL spectrum should be taken into consideration due to the wideband nature of the VLC link. We will elaborate on the related parameters in the Sect. 3.4.2.

## 3.4   System Setup

This section discusses both real-world measurement and synthetic setups to generate the mobility traces and channel dataset.

### 3.4.1   Measurement Setup

Rather than setting the mobility parameters arbitrarily, we have set up a rich body of experiments and measurements on real-world indoor mobility through volunteers' daily lives cell phones to collect traces and used this information to decide the parameters that approximate real mobility as much as possible. However, due to lacking high-resolution detection of light intensity and indoor position using commercial smart phones, we collected the records on UE's orientation and visited locations via an application named "phyphox" [27] from 20 participants who came from different types of offices and laboratories without violating their privacy. This data cannot directly be used for further channel modeling, but they are analyzed to determine the critical parameters in the proposed mobility framework. It should be noted that "phyphox" was used for collecting the records on UE's orientation and visited spot but not for the detailed trajectories since GPS does not present a good accuracy in indoor scenarios. Sometimes indoor GPS errors might exceed 10 m. In this chapter, we restricted several types of offices and laboratories with known layouts. Therefore, we can quantify the accurate location of a visited spot such as a bookshelf based on an approximate location record when the participant dwells there. We measure the distance between each two nodes in a room and categorize the distances into 10 step scale levels for a room. Accordingly, we infer the dimensionless statistics generalized for any indoor layout in any size. The step scale in Fig. 3.4 is therefore obtained by categorizing the distance between two successive identified visited spots. This is why the indoor mobility is still a vital challenge in mobile optical wireless communications (OWC): we cannot get accurate indoor trajectories datasets, and thus we cannot carry out further data-driven investigation accurately. This is however also our motivation: to generate arbitrarily large-size datasets for indoor OWC by reproducing the mobility based on statistics measured from real-world that represents the nature of human behavior.

The first key parameter is $\alpha$ in the bounded Lévy-walk model characterized by Pareto distribution. We analyzed the records on visited spots associated with the step lengths, then estimated an approximation of $\alpha$ as 0.5 for choosing destination distances. As shown in Fig. 3.4, after adopting the same $\alpha$, we get a very similar pattern on the probability distribution of synthetic step scales compared to 307 complete records of measured traces. Note that since the step size in the room is highly relevant to the layout such as the distances among various furnishings, some step scales cannot be reached in a fixed layout. In addition, we determined the parameters for UE's orientations and used the same methodology as in [14] and concluded that for

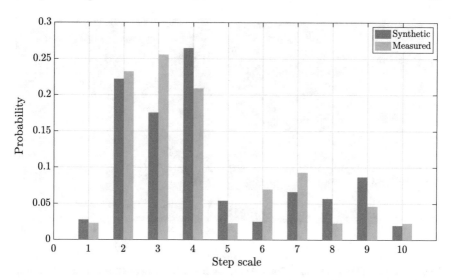

**Fig. 3.4** Statistical analysis of indoor trajectories. All the steps are categorized into several step scales. The step scale distribution of our proposed synthetic trajectory is compared with the measurement results from a similar real-world office room, where $\alpha = 0.5$ for choosing the destination distances [4]

sitting motion: $\mu_s = 45.11$ and $\sigma_s = 7.84$, and for walking motion: $\mu_w = 31.79$ and $\sigma_w = 7.61$, which agree with the results in [14]. The difference might be due to the difference in room layout, user behavior, measurement accuracy, and even amount of data.

## 3.4.2 Synthetic Setup

In this chapter, the room layout and all objects are modeled in the form of cuboid. Specifically, the user body is set as a 1.8 m × 0.2 m × 0.45 m cuboid, where UE is being held at the height of 1.3 m while walking and 1 m while sitting, and at a distance of 0.2 m away from the user's orientation. In the 5 m × 5 m × 3 m room layout shown in Fig. 3.5a, all the desks are set at 1 m length and 0.75 m width with partition height of 1.3 m. In the 10 m × 10 m × 3 m room layout shown in Fig. 3.5b, the desks are set at 2 m length and 1 m width with partition height of 1.3 m. All the cuboids are considered as 5 reflective and opaque surfaces (the bottom surfaces are ignored). Also, the mass of the user is set as 70 kg with a maximum walking speed of 2.1 m/s, and a maximum acceleration of 1 m/s² [28]. The generated mobility sample interval $\delta_\tau$ is 100 ms. The trajectory duration determined by return regularity follows a Gaussian distribution with a mean of 1,000 samples and a standard deviation of 50. For the sojourn duration at each resident node, we assume $\beta = 1$ and randomly select $t_{s,\max}$ and $t_{s,\min}$ for each furniture. However, this chapter only reflects movement

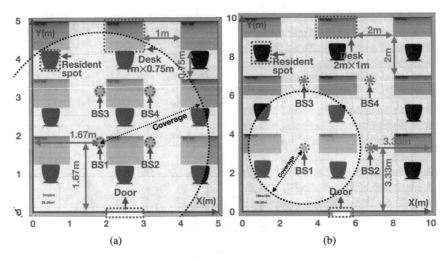

**Fig. 3.5** **a** Room layout of 5 m × 5 m × 3 m setup. **b** Room layout of 10 m × 10 m × 3 m setup. Four light BSs are distributed evenly on the ceiling plane [4]

impacts, which means the detailed model of sojourn behavior is not in the scope of our investigation. All the statistics we derive for the optical channel are deemed during users' mobility without sojourn state since the existing literature has given plenty of results on OWC under stationary state. Interested readers still could synthesize the complete trajectory data including sojourn state using our framework.

According to the procedures we have introduced in the preceding sections, we produce the human trajectories dataset as demonstrated in Fig. 3.6a, c. In Fig. 3.6b, d, we illustrate the spatial distribution of trajectories in which the frequency of each unit area being visited is calculated, and we only show the case of moving. Figure 3.6b, d demonstrate a higher probability of passing by the resident nodes than the crossing paths, and a higher probability in the resident nodes and path located along the central row than other rows, and a higher probability around turning corners than typical paths.

Table 3.1 summarizes the transceiver parameters used following most of the existing literature [9, 13] while generating the channel datasets. We consider an OWC system where the downlink is supported by VLC and the uplink is supported by IR. When generating the channel information, we set the UE transceivers located at the center of the top of the phone pointing to the azimuth direction. Four light BSs are located evenly on the ceiling as illustrated in Figs. 3.5a, b. Orthogonal bands are allocated to the BSs. We consider the transmit power of each BS as a dimensionless unit power, because we only focus on the channel gain. The time resolution for CIR is 0.1 ns. In this chapter, for simplicity, we assume a certain light spectral power distribution of visible light transmitter such that the resulting reflectance of visible light can be treated as in Table 3.1. Readers can adopt more detailed spectral reflectance data based on the actual measurements if necessary.

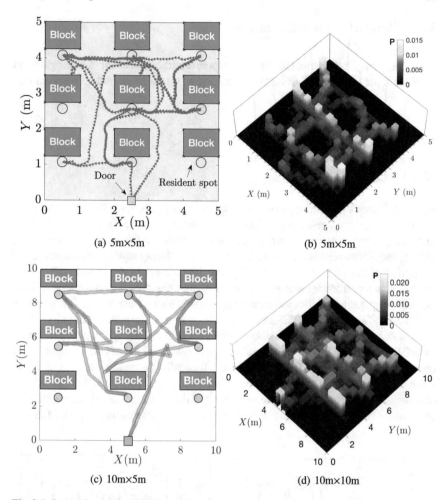

Fig. 3.6 In **a** and **c**, an example of reproduced trajectory is given. In **b** and **d**, the frequency of each unit area being visited is calculated from the generated dataset [4]

**Table 3.1** Parameters for transceivers

| BS number | 4 | $\Phi_{1/2}$ | 60° |
|---|---|---|---|
| $A_R$ | 100 mm² | $\Psi$ | 45° |
| VLC wavelength | 380–780 nm | VLC $\rho_\varepsilon$ | 0.7 (walls), 0.2 (other surfaces) |
| IR wavelength | 800 nm | IR $\rho_\varepsilon$ | 0.8 (walls), 0.5 (other surfaces) |
| $\varepsilon$ area | $10^4$ mm² | $\{\varepsilon\}$ model | Lambert |

## 3.5   Channel Characterization

In the following, we discuss spatial and temporal characteristics for channel gain and
bandwidth for LOS and NLOS components, based on the generated channel dataset.

### 3.5.1   Overall Statistics

#### 3.5.1.1   Channel Gain

Figure 3.7 shows a glimpse of one complete record on channel gain along with the
complete trajectory shown in Fig. 3.6( a. Note that in order to show the whole case,
including sitting motions, we did not delete the records on the sojourn. However, for
sake of mobility analysis, we re-scaled the records during sojourn states, shrinking
the period of the sojourn in a resident spot to 1000 times shorter as in Fig. 3.7. Furthermore, in the following statistics, due to the same reason, we excluded the records
on stationary status and focused on moving periods exclusively. In Fig. 3.7, one may
find the relevance between different BSs owing to the geometric symmetry of the
room layout. Sometimes this relevance shows up as similarity on channel gains such
as in BS 1 and BS 3. However, sometimes it emerges as a kind of complementarity if
one had noticed BS 1 and BS 2 for instance. Basically, the channel gain in uplink is
close to that in downlink; however, the uplink holds an obviously higher probability
of outage. In this chapter, LOS outages are defined as complete occlusions in LOS
rays, while NLOS outages are referred to as complete occlusions in both LOS and

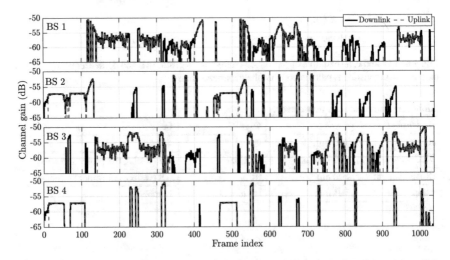

**Fig. 3.7** Illustration of the LOS channel gain linked with 4 light BSs for both uplink and downlink
during an entire trajectory, where the sample interval is 100 ms [4]

NLOS rays. Such occlusions can be induced by being either blocked by any opaque object or out of the FOV of the receiver.

Next, we examine the overall statistics of the channel gain. From Fig. 3.8a, it is shown that the PDFs exhibit multiple peaks for LOS links. When the UE is moving, it experiences different relative positions against the BSs in terms of orientation and distance, and thus the distribution of the channel gain is associated with the mobility pattern. As aforementioned, the users pass by some specific areas with higher probabilities; therefore, several peaks emerge in the PDF of the channel gain, which feature the regularity of the channel status at those frequently visited spots. For the PDFs shown in Fig. 3.8a, the channel gain $\hat{H}_0$ distribution is fitted by $N_{\text{LOS}}$ parts of Nakagami distribution with respective weights. The overall PDF of multiple-peak Nakagami distribution in LOS links is given by $f(\hat{H}_0) = f_{\text{MN}}(\hat{H}_0; k_n, \mu_n, \omega_n)$ where

$$f_{\text{MN}}(\hat{H}_0; k_n, \mu_n, \omega_n) = \sum_{n=1}^{N_{\text{LOS}}} k_n \frac{2\mu_n^{\mu_n} \hat{H}_0^{2\mu_n - 1}}{\Gamma(\mu_n)\omega_n^{\mu_n}} e^{\left(-\frac{\mu_n}{\omega_n}\hat{H}_0^2\right)}, \tag{3.22}$$

$\Gamma(\mu_n)$ stands for the Gamma function, $k_n$ represents the weights ($\sum k_n = 1$), $\mu_n$ stands for the shape parameter, and $\omega_n$ denotes the scale parameter. The detailed parameters of LOS channel gain distribution are listed in Table 3.2.

On the other hand, for the 1st reflection representing the NLOS component shown in Fig. 3.8b, the channel gain $\hat{H}_1$ without outages, however, adheres to conventional distributions. For the downlinks, the generalized log-logistic distributions (Burr Type XII distribution) can be applied as $f(\hat{H}_1) = f_{\text{B}}(\hat{H}_1; c_B, k_B, \lambda_B)$ where

$$f_{\text{B}}(\hat{H}_1; c_B, k_B, \lambda_B) = \frac{c_B k_B}{\lambda_B} \left(\frac{\hat{H}_1}{\lambda_B}\right)^{c_B - 1} \left[1 + \left(\frac{\hat{H}_1}{\lambda_B}\right)^{c_B}\right]^{-k_B - 1}, \tag{3.23}$$

and $c_B$, $k_B$, and $\lambda_B$ correspond to the first and second shape parameters, and the scale parameter, respectively. However, for the uplinks in the first reflection NLOS links, the distributions for different scenarios diverged as $\hat{H}_1$ for the $10\,\text{m} \times 10\,\text{m} \times 3\,\text{m}$ layout is subject to a Gamma distribution:

$$f_{\text{G}}(\hat{H}_1; a_G, b_G) = \frac{\hat{H}_1^{a_G - 1} e^{\frac{-\hat{H}_1}{b_G}}}{b_G^{a_G} \Gamma(a_G)}, \tag{3.24}$$

where $a_G$ and $b_G$ denote the parameters of shape and scale, respectively; and for the NLOS uplinks in $5\,\text{m} \times 5\,\text{m} \times 3\,\text{m}$ layout, $\hat{H}_1$ is subject to a Nakagami distribution:

$$f_{\text{N}}(\hat{H}_1; \mu, \omega) = \frac{2\mu^{\mu} \hat{H}_1^{2\mu - 1}}{\Gamma(\mu)\omega^{\mu}} e^{\left(-\frac{\mu}{\omega}\hat{H}_1^2\right)}, \tag{3.25}$$

where the detailed parameters of NLOS channel gain distribution are listed in Table 3.3.

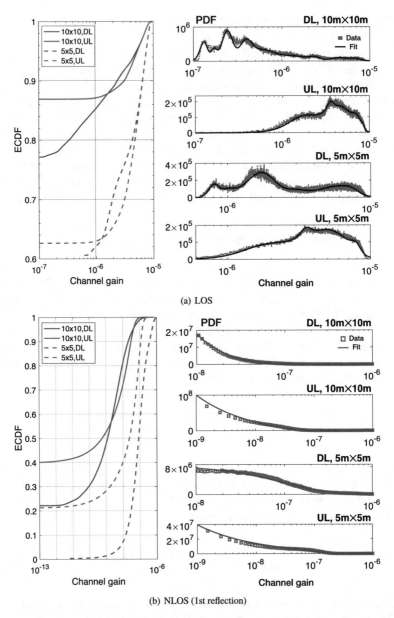

(a) LOS

(b) NLOS (1st reflection)

**Fig. 3.8** Distributions of channel gain through all light links. The empirical cumulative distribution functions (ECDFs) illustrate overall statistics including outages, whereas the PDFs indicate the statistics and corresponding distribution fitting without outages. In (**a**), the multiple-peak Nakagami distributions are shown in the LOS PDFs for all LOS links. In (**b**), the channel gain statistics of the 1st reflection representing the major component in NLOS are shown, where their uplink distributions without outages are subject to Gamma and Nakagami distributions, and for downlinks to Burr distributions [4]

**Table 3.2** LOS channel gain distribution fit parameters (minimal data unit: $10^{-10}$)

UL, $10\,\text{m} \times 10\,\text{m} \times 3\,\text{m}$

| $n$ | $k_n$ | $\mu_n$ | $\omega_n$ |
|---|---|---|---|
| 1 | 0.09 | 2.79 | $3.14 \times 10^8$ |
| 2 | 0.75 | 1.93 | $2.47 \times 10^9$ |
| 3 | 0.03 | 40.43 | $1.25 \times 10^9$ |
| 4 | 0.015 | $1.82 \times 10^2$ | $7.13 \times 10^9$ |
| 5 | 0.115 | 21.28 | $5.31 \times 10^9$ |

UL, $5\,\text{m} \times 5\,\text{m} \times 3\,\text{m}$

| $n$ | $k_n$ | $\mu_n$ | $\omega_n$ |
|---|---|---|---|
| 1 | 0.06 | 3.10 | $3.38 \times 10^8$ |
| 2 | 0.03 | 38.97 | $1.25 \times 10^9$ |
| 3 | 0.79 | 2.19 | $3.14 \times 10^9$ |
| 4 | 0.06 | 46.21 | $6.28 \times 10^9$ |
| 5 | 0.06 | 2.36 | $8.32 \times 10^8$ |

DL, $10\,\text{m} \times 10\,\text{m} \times 3\,\text{m}$

| $n$ | $k_n$ | $\mu_n$ | $\omega_n$ |
|---|---|---|---|
| 1 | $1.60 \times 10^{-2}$ | 28.38 | $1.75 \times 10^6$ |
| 2 | $6.10 \times 10^{-2}$ | 15.03 | $5.99 \times 10^6$ |
| 3 | $6.10 \times 10^{-2}$ | 12.46 | $1.59 \times 10^7$ |
| 4 | $8.00 \times 10^{-3}$ | 53.29 | $2.92 \times 10^6$ |
| 5 | $3.00 \times 10^{-1}$ | 0.96 | $2.45 \times 10^8$ |
| 6 | $5.00 \times 10^{-2}$ | 18.59 | $5.72 \times 10^8$ |
| 7 | $2.00 \times 10^{-1}$ | 6.84 | $2.07 \times 10^9$ |
| 8 | $2.00 \times 10^{-1}$ | 10.44 | $4.65 \times 10^9$ |
| 9 | $1.10 \times 10^{-1}$ | 2.36 | $3.72 \times 10^7$ |

DL, $5\,\text{m} \times 5\,\text{m} \times 3\,\text{m}$

| $n$ | $k_n$ | $\mu_n$ | $\omega_n$ |
|---|---|---|---|
| 1 | $2.00 \times 10^{-2}$ | 42.80 | $6.42 \times 10^7$ |
| 2 | $2.60 \times 10^{-1}$ | 5.60 | $3.16 \times 10^8$ |
| 3 | $7.00 \times 10^{-2}$ | 19.36 | $5.47 \times 10^9$ |
| 4 | $1.10 \times 10^{-1}$ | 5.80 | $8.61 \times 10^8$ |
| 5 | $2.20 \times 10^{-2}$ | 17.22 | $9.94 \times 10^7$ |
| 6 | $5.18 \times 10^{-1}$ | 3.49 | $3.84 \times 10^9$ |

**Table 3.3** NLOS channel gain distribution fit parameters (minimal data unit: $10^{-10}$)

| UL, $10\,\mathrm{m} \times 10\,\mathrm{m} \times 3\,\mathrm{m}$ | | |
|---|---|---|
| $a_G$ | $b_G$ | |
| $4.99 \times 10^{-1}$ | $3.00 \times 10^2$ | |
| UL, $5\,\mathrm{m} \times 5\,\mathrm{m} \times 3\,\mathrm{m}$ | | |
| $\mu$ | $\omega$ | |
| $2.77 \times 10^{-1}$ | $5.65 \times 10^5$ | |
| DL, $10\,\mathrm{m} \times 10\,\mathrm{m} \times 3\,\mathrm{m}$ | | |
| $\lambda_B$ | $c_B$ | $k_B$ |
| $1.13 \times 10^3$ | $5.83 \times 10^{-1}$ | 6.49 |
| DL, $5\,\mathrm{m} \times 5\,\mathrm{m} \times 3\,\mathrm{m}$ | | |
| $\lambda_B$ | $c_B$ | $k_B$ |
| $6.71 \times 10^3$ | $9.90 \times 10^{-1}$ | 5.53 |

### 3.5.1.2 Bandwidth

In view of the high probability of outages in LOS transmissions, the NLOS standalone channels should also be considered, thereby increasing the flexibility and stability of deploying Li-Fi in mobile scenarios. The connectivity of NLOS standalone links is much higher as shown in Fig. 3.8, and NLOS links could maintain communications to a certain extent when LOS links are blocked. However, due to multi-path effect in NLOS links, the modulation bandwidth is limited, and therefore deserves a special investigation.

As illustrated in Fig. 3.9, we fit the PDF of the uplink bandwidth $\hat{B}$ using the generalized log-logistic distribution as $f(\hat{B}) = f_{\mathrm{B}}(\hat{B}; c_B, k_B, \lambda_B)$

$$f_{\mathrm{B}}(\hat{B}; c_B, k_B, \lambda_B) = \frac{c_B k_B}{\lambda_B} \left( \frac{\hat{B}}{\lambda_B} \right)^{c_B - 1} \left[ 1 + \left( \frac{\hat{B}}{\lambda_B} \right)^{c_B} \right]^{-k_B - 1}. \tag{3.26}$$

However, for downlinks, the bandwidth follows multiple-peak log-logistic distributions as $f(\hat{B}) = f_{\mathrm{ML}}(\hat{B}; k_n, \mu_n, \sigma_n)$, where

$$f_{\mathrm{ML}}(\hat{B}; k_n, \mu_n, \sigma_n) = \sum_{n=1}^{N_{\mathrm{BW}}} k_n \frac{e^{z_n}}{\sigma_n \hat{B} (1 + e^{z_n})^2}, \tag{3.27}$$

and $z_n = \frac{\log(\hat{B}) - \mu_n}{\sigma_n}$, $\mu_n$ and $\sigma_n$ correspond to the mean of logarithmic values and the scale parameter of logarithmic values, respectively, for the $n$-th peak. The detailed parameters of NLOS bandwidth distribution are listed in Table 3.4.

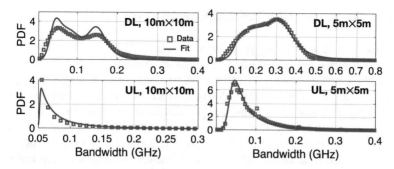

**Fig. 3.9** PDF of −3 dB modulation bandwidth (BW) through all optical NLOS standalone links, where the fitting curves are also shown. The uplink bandwidth follows generalized log-logistic distributions and the downlink bandwidth follows multiple-peak log-logistic distributions [4]

**Table 3.4** Bandwidth distribution fit parameters (minimal data unit: 0.1 Hz)

| UL, 10 m × 10 m × 3 m | | | |
|---|---|---|---|
| $\lambda_B$ | $c_B$ | $k_B$ | |
| $5.19 \times 10^8$ | $7.02 \times 10^1$ | $2.92 \times 10^{-2}$ | |
| **UL, 5 m × 5 m × 3 m** | | | |
| $\lambda_B$ | $c_B$ | $k_B$ | |
| $8.28 \times 10^8$ | 5.34 | $2.62 \times 10^{-1}$ | |
| **DL, 10 m × 10 m × 3 m** | | | |
| $n$ | $k_n$ | $\mu_n$ | $\sigma_n$ |
| 1 | 0.3 | 20.7 | $2.08 \times 10^{-1}$ |
| 2 | 0.35 | 21.8 | $9.92 \times 10^{-2}$ |
| 3 | 0.35 | 21.3 | $2.03 \times 10^{-1}$ |
| **DL, 5 m × 5 m × 3 m** | | | |
| $n$ | $k_n$ | $\mu_n$ | $\sigma_n$ |
| 1 | 0.1 | 21.0 | $1.89 \times 10^{-1}$ |
| 2 | 0.3 | 21.9 | $1.13 \times 10^{-1}$ |
| 3 | 0.6 | 21.6 | $2.75 \times 10^{-1}$ |

### 3.5.1.3 Statistical Divergence

Such a divergence between LOS and NLOS and different room layouts imply a close entanglement of the channel statistics with environment-confined mobility details spatially and temporally. The multi-peak probability distribution in LOS channel gain clearly expresses the influence of the trajectory-environment interaction pattern on the channel gain distribution. However, such interaction pattern is not noticeable through NLOS statistics of channel gain or bandwidth. Thus, we can no longer use a single statistical model to summarize the indoor optical channel. This inspires us to explore further the temporal and spatial characteristics behind these distributions.

#### 3.5.1.4    Comparison with Random Waypoint-Based Model

The conventional random waypoint-based channel statistic introduced in [13] suggests that the channel gain follows Rayleigh distribution. However, our statistics show that the outage probability is higher during movement of UE, and the pattern of channel gain distribution highly depends on the environment-confined mobility pattern, which cannot be modeled using a simple (single) stochastic model. The most recent statistics in [29] consider random orientation of UE and suggest that for stationary users, the channel gains follow the modified truncated Laplace model and the modified Beta model, while for mobile users, the channel gains follow the sum of modified truncated Gaussian model and the sum of modified Beta model. However, since the user mobility in [29] is assumed to follow the random waypoint model, likewise, its user trajectory distribution cannot reflect the influences from the frequency and periodic pattern of visited locations according to the macro mobility pattern, i.e., when to go and where to go. Different spatio-temporal trajectory proportions yield different channel statistics, and this is why our statistics differ from the random waypoint-based models after taking into account the nature of human behavior. More importantly, using the framework presented in this chapter, we can elaborate deeply on more details featuring the spatio-temporal evolution of channel statistics, as will be highlighted next.

### 3.5.2  Spatial Features

Next, we investigate the spatial characteristics of the outage probability. The spatial distributions of LOS outage probability in $10\,m \times 10\,m$ and $5\,m \times 5\,m$ room layouts are illustrated in Fig. 3.10a–e. The results for both uplink and downlink confirm the higher probability of outage at the intersections around the entrance as well as the resident nodes apart from the central area. This is due to the fact that the steering motion of user prompts turbulence in orientation. Because of the asymmetry in coverage, the distribution patterns in both directions are not identical as the uplinks show a steeper gradient in spatially outage probability changes.

Then, we examine the spatial features in ECDFs of LOS channel gain, as depicted in Fig. 3.10c, f. Although the ECDFs in different regions present apparent differences, they generally exhibit geometric symmetry. This geometrical feature is inherited from the symmetry of the furnishings and BS layout. These results illustrate a better channel state in the middle, but more severe features at the edges. The areas with better channel conditions contain the most aisles. In particular, the sub-figures in the middle parallel to the horizontal direction (Row C: $y \in (4m, 6m)$ in $10\,m \times 10\,m$ and $y \in (2m, 2.75\,m)$ in $5\,m \times 5\,m$) represent the area where users tend to pass by with a higher probability. Nevertheless, the uplink still outperforms the downlink slightly with a higher mean channel gain once there exists a received signal (also can be noticed from Fig. 3.8), and this fact is due to the difference in incident angle between transmitters and receivers.

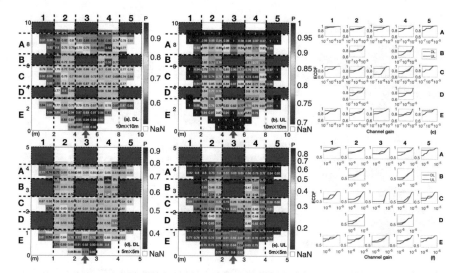

**Fig. 3.10** Spatial outage probabilities and ECDFs of LOS channel gain. ECDFs are carried out outside furnishings area indexed through A to E on a vertical axis and through 1–5 on a horizontal axis. The black blocks ■ denote the furnishings areas and the red arrows ↑ stand for the entrances [4]

However, the spatial patterns in NLOS links as shown in Figs. 3.11a–e exhibit different behavior. For uplinks, we find more outages around the entrance area, but we notice there are fewer outages around the entrance in downlink. The remaining areas demonstrate a stationary spatial pattern since the blockages and FOV cannot totally conceal all the diffused rays. This also limits the spatial variation range in bandwidth statistics since the key point to limiting bandwidth is the received ray delay spread, where the arrival time of a ray depends on the light path length rather than the relative angle between the transceivers.

It should be highlighted that the spatial statistics are not sufficient to understand the performance of indoor optical mobile channels. We ought to introduce the temporal statistics that are collected within a specific time window, during which the user trajectory follows the distribution illustrated in Fig. 3.6b, d with respect to the time interval. When evaluating the overall performance, the spatial features alone are not enough since the frequency of visiting a specific area is not taken into account. However, in the temporal statistics, the average outage probability during a certain time interval is a combination of all possible spatial distributions with the corresponding weights (frequency of visits). Thus, the temporal and spatial behaviors appear to be different. This motivates the need for a deeper investigation of the temporal characteristics, a study which will be carried out in the next subsection.

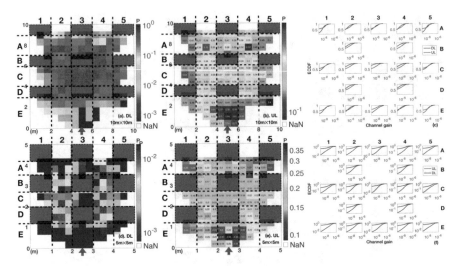

**Fig. 3.11** Spatial outage probabilities and ECDFs of NLOS channel gain. ECDFs are carried out outside furnishings area indexed through A to E on a vertical axis and through 1–5 on a horizontal axis. The black blocks ■ denote the furnishings areas and the red arrows ↑ stand for the entrances [4]

### 3.5.3 Temporal Features

Figure 3.12 presents the channel gain and outage probability for the complete temporal structure of the mobility pattern, where time indices of different traces are unified and categorized into 1,000 intervals, and then the channel statistics are conducted, respectively. The temporal patterns and their dependence on trajectories are illustrated by an explicit exposure on the channel statistics evolution throughout the motions of entering, wandering and exiting. Additionally, Fig. 3.13 illustrates the impact of mobility on bandwidth in NLOS links. We discuss the results according to the following stages:

#### 3.5.3.1 Stage 1: Room Entering

For the LOS component, the temporal evolution of the extreme channel states while moving from the entrance to the first destination is revealed at Stage 1, namely, the room entering stage. For the uplink in both room layouts, the outage probabilities begin with 100%, then fall sharply in a volatile manner, then slightly rise up, and afterward wanders around 60% in the 5 m×5 m room layout and 85% in the 10 m×10 m room layout. However, for the downlink, the outage probability in the 5 m×5 m layout begins with around 50% and then fluctuates in a broad range of 50% until it fluctuates around 60%, which is sightly lower than in the uplink. Whereas in the 10 m × 10 m layout, the outage probability starts around 75%, then it also fluctuates in a broad range of 40% until fluctuating around 70%. As for the NLOS, the

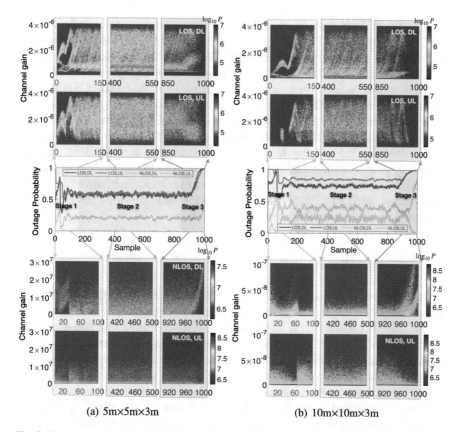

**Fig. 3.12** Temporal outage probability and PDF of channel gain with respect to time along the course of trajectories. Time indices of different traces are unified and then classified into 1,000 intervals in which outage statistics are carried out, respectively. Stages 1–3 correspond to the three time-blocks during the entering, wandering, and exiting movements [4]

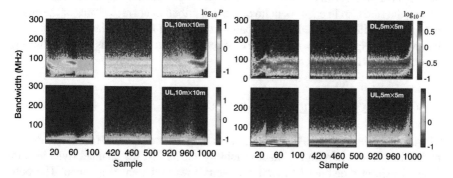

**Fig. 3.13** PDFs of −3 dB modulation bandwidth with respect to time along the course of 3 stages [4]

outage probability is much lower for the first reflection component than the LOS. In the $5\,\mathrm{m} \times 5\,\mathrm{m}$ room layout, the outage probability can hold below 1% in the downlink, and exhibits a similar behavior in the uplink as the LOS downlink but with outage probability around 20%. In the $10\,\mathrm{m} \times 10\,\mathrm{m}$ layout, the outage probability of NLOS goes higher with broader dynamic range than that of LOS. As shown in Fig. 3.12, the PDFs of channel gain with respect to time along the course of Stage 1 is equivalent to an expansion of Fig. 3.8 in a successive temporal order. The influence of entering motion is very obvious. The peaks of the channel gain PDFs for the LOS component are shifting over time with the entering movement; whereas in NLOS links, although there is no prominent distribution peak, the shape of the PDF also changes with the entering movement. In this stage, as depicted in Fig. 3.13, the bandwidth distributions are not stable either, where the peaks vibrate while the PDF curves gradually broaden until the user reaches the first destinations.

### 3.5.3.2  Stage 2: Room Wandering

After 100 intervals from the beginning of the trajectory, the outage probabilities become stationary random processes as denoted by Stage 2, namely, the room wandering stage. When the user is wandering within the room and shuttling among the resident nodes, the movement period is defined as a wandering stage. The outage probability curves for uplink and downlink undergo consistent oscillations in Stage 2. Besides, the average outage probability in terms of time shown in Fig. 3.12 is higher than the outage probabilities presented from the spatial perspective of many areas as demonstrated in Fig. 3.10 since the areas with a high probability of outage are spatially small, but they are frequently passed by from the perspective of time. Meanwhile, the distribution characteristics are relatively stable. That is to say, from a time perspective, in addition to Stage 1 and Stage 3 (as will be interpreted later), the channel gain sequence can be seen as a stationary stochastic process with a multi-peak hybrid distribution. However, from a spatial perspective, given the differences among the statistics in each room layout, the distribution characteristics of the channel gain are still dominated by the spatial structure of the room layout. As for the bandwidth in Fig. 3.13, the PDF curves are becoming more stable as the user shuttles between different resident nodes.

### 3.5.3.3  Stage 3: Room Exiting

The last stage is defined as the motion of heading towards the exit (entrance) driven by the return tendency in the macro mobility. Generally, the channel dynamics in Stage 3 exhibit a distinct temporal symmetry toward the entering stage. The outage probabilities soar towards 100% before leaving the room except for the NLOS downlinks. In the PDFs of LOS channel gain, we find the peaks reciprocating among the distribution region over time; but the peaks of the PDFs of NLOS channel gain seem to shift to higher gain over time in this stage. The distribution peak diversion towards higher values emerging at the end of Stage 3 is also found in the PDFs of bandwidth apart from the uplink in $10\,\text{m} \times 10\,\text{m}$ room layout. Such peak diversions in the NLOS downlinks reflect the improvement in channel status with the outage rate's diminishing and the distribution peaks' rising in channel gain and bandwidth. Yet in other links, the channels are deteriorating along with the exiting movement due to the increase in outage occasions.

### 3.5.4 Handover Rate

Handovers are triggered among BSs to ascertain reliable connection with quality-of-service guarantee. We assume the handover process overhead as 300 ms [1] and the BS with the best LOS channel gain will be designated as the target BS. During handover processes, the UEs dwell in the currently serving BS. The handover rate under the indoor mobility model is shown in Fig. 3.14. The means of handover rate are 0.79, 0.85, 0.24, 0.76 for the downlinks in $10\,\text{m} \times 10\,\text{m}$ room layout and $5\,\text{m} \times 5\,\text{m}$ room layout, and the uplinks in $10\,\text{m} \times 10\,\text{m}$ room layout and $5\,\text{m} \times 5\,\text{m}$ room layout, respectively. We note that the handover rate of uplink in the $10\,\text{m} \times 10\,\text{m}$ layout is actually not proportional to the outage risk. The reason of this fact is that the data stream has nowhere to be offloaded since all BSs are suffering poor links under such high outage risks, which in turn undermines the handover rate. Of course, this ill-conditioned result is caused by poor link management. Therefore, in view of human indoor mobility characteristics, a more intelligent methodology on network resource management is needed. This important topic is addressed in the next chapter.

**Fig. 3.14** Handover rate under user's indoor mobility [4]

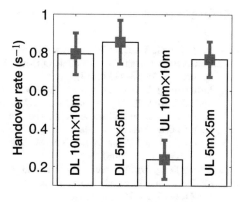

## 3.6   Summary

In order to address the limitations of the existing models and hence offer an accurate tool for proper characterization and realization of wireless channels at high frequency bands, this chapter carried out the following tasks:

- A mobility model that better captures the nature of human mobility was proposed. The presented model exhibits a macro pattern featured by a semi-Markov renewal process subject to Lévy-walk and return regularity, as well as a micro pattern featured by the shortest path, steering behavior, and random UE orientations.
- Based on the proposed mobility model, a systematic framework was put forward to generate mobile trajectories and to generate CIR datasets for indoor channels. Key parameters of this framework are obtained from fitting measurement statistics in real-world from anonymous mobility collections.
- The developed framework was employed to characterize the channel of indoor mobile optical wireless networks with VLC support in the downlink and infrared communications support in the uplink, at practical time and space scales and in the presence of blockages from furniture and user body. The proposed framework can be further adopted to characterize indoor wireless channels at high frequency bands such as mmWave and Terahertz once appropriate channel models are employed.
- The statistical properties of mobile optical wireless channels were analyzed from both time and space perspectives. Space-time-dependent statistical patterns for channel gain, bandwidth, and outage probability were proposed, and which dramatically evolve over the environment-confined mobility.

## References

1. H. Haas et al., What is LiFi? J. Lightw. Technol. **34**, 1533–1544 (2016)
2. Harald Haas, Cheng Chen, Dominic O'Brien, A guide to wireless networking by light. Prog. Q. Electron. **55**, 88–111 (2017)
3. J. Armstrong, Y.A. Sekercioglu, A. Neild, Visible light positioning: a roadmap for international standardization. IEEE Commun. Mag. **51**(12), 68–73 (2013)
4. Z. Wu et al., Channel characterization and realization of mobile optical wireless communications, in *IEEE Transactions on Communications 2020* (under review)
5. Ivan Moreno and Ching-Cherng Sun, Modeling the radiation pattern of LEDs. Opt. Express **16**(3), 1808–1819 (2008)
6. Jupeng Ding, Xu Zhengyuan, Lajos Hanzo, Accuracy of the point source model of a multi-LED array in high-speed visible light communication channel characterization. IEEE Photon. J. **7**(4), 1–14 (2015)
7. J.M. Kahn, J.R. Barry, Wireless infrared communications, in *Proceedings of the IEEE 85.2* (1997), pp. 265–298. ISSN: 0018-9219
8. Henrik Schulze, Frequency-domain simulation of the indoor wireless optical communication channel. IEEE Trans. Commun. **64**(6), 2551–2562 (2016)
9. K. Lee, H. Park, J.R. Barry, Indoor channel characteristics for visible light communications. IEEE Commun. Lett. **15**(2), 217–219 (2011). ISSN: 1089-7798

10. Volker Jungnickel et al., A physical model of the wireless infrared communication channel. IEEE J. Sel. Areas Commun. **20**(3), 631–640 (2002)
11. Cheng Chen et al., Efficient analytical calculation of non-line-of-sight channel impulse response in visible light communications. J. Lightwave Technol. **36**(9), 1666–1682 (2017)
12. Farshad Miramirkhani et al., A mobile channel model for VLC and application to adaptive system design. IEEE Commun. Lett. **21**(5), 1035–1038 (2017)
13. Petr Chvojka et al., Channel characteristics of visible light communications within dynamic indoor environment. J. Lightwave Technol. **33**(9), 1719–1725 (2015)
14. M.D. Soltani et al., Modeling the random orientation of mobile devices: Measurement, analysis and LiFi use case. IEEE Tran. Commun. **67**(3), 2157–2172 (2018)
15. A. Sewaiwar, S.V. Tiwari, Y.H. Chung, Mobility support for full-duplex multiuser bidirectional VLC networks. IEEE Photon. J. **7**(6), 1–9 (2015)
16. I. Rhee et al., On the Levy-walk nature of human mobility. IEEE/ACM Trans. Netw. **19**(3), 630–643 (2011). ISSN: 1063-6692
17. Gonzalez, M.C., Hidalgo, C.A., Barabasi, A-L.: Understanding individual human mobility patterns. Nature **453**(7196), 779 (2008)
18. E.W. Dijkstra, A note on two problems in connexion with graphs. Numer. Math. **1**(1), 269–271 (1959)
19. C.W. Reynolds, Steering behaviors for autonomous characters, in *Game Developers Conference*, vol. 1999, Citeseer (1999), pp. 763–782
20. H. Farooq, A. Imran, Spatiotemporal mobility prediction in proactive self-organizing cellular networks. IEEE Commun. Lett. **21**(2), 370–373 (2017). ISSN: 1089-7798
21. A. Nadembega, A. Hafid, T. Taleb, A destination and mobility path prediction scheme for mobile networks. IEEE Trans. Veh. Technol. **64**(6), 2577–2590 (2015). ISSN: 0018-9545
22. A.A. Purwita et al., Terminal orientation in OFDM-based LiFi systems. IEEE Trans. Wirel. Commun. 1–1 (2019). ISSN: 1536-1276
23. Z. Wu et al., Efficient prediction of link outage in mobile optical wireless communication, in *IEEE Transactions on Wireless Communications 2020* (under review)
24. John R. Barry et al., Simulation of multipath impulse response for indoor wireless optical channels. IEEE J. Sel. Areas Commun. **11**(3), 367–379 (1993)
25. S.P. Rodriguez et al., Simulation of impulse response for indoor visible light communications using 3D CAD models. EURASIP J. Wirel. Commun. Netw. **2013**(1), 7 (2013)
26. Lubin Zeng et al., High data rate multiple input multiple output (MIMO) optical wireless communications using white LED lighting. IEEE J. Sel. Areas Commun. **27**(9), 1654–1662 (2009)
27. https://phyphox.org/
28. B.J. Mohler et al., Visual flow influences gait transition speed and preferred walking speed. Exp. Brain Res. **181**(2), 221–228 (2007)
29. M.A. Arfaoui et al., Measurements-based channel models for indoor LiFi systems (2020), arXiv:2001.09596

# Chapter 4
# Data-driven Handover Framework in Mobile 5G+ HetNets

**Abstract** This chapter investigates resource management in 5G and beyond (5G+) heterogeneous networks (HetNets) for multi-mode connections. To ensure a reliable link with quality-of-service (QoS) guarantee in terms of network delay and through-put, vertical handovers are triggered within the HetNet. A cross-layer data-driven approach is adopted to reach optimal handover decisions and tackle the challenges associated with mobility and reliability. We consider the application scenario of a HetNet consisting of an overlapped coverage of radio frequency (RF) and optical wireless communication (OWC) to support human connectivity in the 5G+ indoor HetNets. Using the dataset presented in the previous chapter, we present a data-driven algorithm that predicts abrupt outages in line-of-sight (LOS) optical links and evalu-ates the optical channel quality through deep learning. Given the resulting LOS link outage prediction in OWC, a reinforcement learning-based approach is introduced to implement optimal vertical handover decisions with QoS guarantee. This handover decision algorithm learns to make a trade-off between the outage risk and the cost of excessive handovers. Numerical results demonstrate considerable improvement in overall latency and handover rate under indoor mobility for bi-directional links.

## 4.1 Introduction

With the rapid development of 5G and beyond (5G+) communications, the wire-less medium has evolved into a complex heterogeneous network (HetNet) [1] that consists of multiple networks equipped with different communication technologies that present overlapped coverage with different cell sizes. In such an environment, cooperation among overlapped networks to ensure quality-of-service (QoS) support for mobile users is deemed imperative. This chapter investigates one application sce-nario of a 5G+ HetNet that integrates optical links (for high throughput) and radio frequency (RF) links (for high link reliability) to support user's QoS requirements [2]. In this RF-optical HetNet, two types of base stations (BSs) are considered, namely, macro and small BSs [3]. The small BS features optical links such as visible light communication (VLC) for downlink and infrared (IR) for uplink, both with high

Z.-Y. Wu et al., *Efficient Integration of 5G and Beyond Heterogeneous Networks*,
https://doi.org/10.1007/978-981-15-6938-8_4

capacity but also with a high risk of abrupt link outages in the presence of human mobility, as discussed in the previous chapter. The macro BS working in the RF spectrum offers a more stable and wider coverage but with lower throughput. Therefore, in a practical mobile scenario, the macro BSs are used to address the frequent instantaneous vertical handovers [4], which result in a network delay problem owing to the presence of unnecessary excessive handovers.

## 4.1.1 Background

### 4.1.1.1 Handovers

Recent studies have laid a solid foundation through various solutions and insights to tackle vertical handovers in RF-optical HetNets. In [5], a fuzzy logic system is developed to manage the BS selection problem in RF-optical HetNets. In [6], a handover model is introduced by taking into account the random waypoint (RWP) model for movement of user equipment (UE) and a geometric model for UE's orientation. By utilizing the effective capacity [7] as the link-layer channel model, resource allocation under QoS constraints has been studied in an RF-optical HetNet for both multi-homing and multi-mode scenarios [8]. In addition to the effective capacity, implementing a two-state discrete-time Markov process as an ON-OFF data source [9] has been found very useful in modeling the vertical handover problem. Besides, in [10], a dynamic approach is adopted to obtain a trade-off between switching (handover) cost and delay requirement. A similar approach has been introduced to manage vertical handovers in millimeter wave (mmWave) networks [11], where by utilizing the user's mobility information (velocity and location), the mechanism in [11] avoids excessive handovers and effectively tackles beamforming misalignment and signal blockages in mmWave small cells.

### 4.1.1.2 Challenges

As highlighted, the existing research works on handover policy design rely on a simple random walk and stochastic process assumptions, which are inappropriate for modeling blockage-sensitive optical wireless networks subject to user's mobility. Moreover, taking advantage of indoor location information is still not realistic, because, at present, the accuracy of indoor mobile positioning is low, which will further mislead the network resource management. Since mobility patterns are essential to model and investigate a mobile network, the following two important questions have to be answered: how a practical RF-optical HetNet may perform along realistic indoor user trajectories, and whether the channel fluctuations can be predicted and leveraged to pursue an optimal network management policy. Also, although the emerging deep learning approaches have shown effectiveness in dealing with handover management in RF standalone networks [12], the smart handover and link assignment strategies in 5G+ HetNet are more challenging due to the distinguished scopes of mobility and various air-interface features.

Overall, the RF-optical air-interfaces in UEs should enable mobile human connectivity in 5G+ HetNets while ensuring a seamless smart handover policy with QoS guarantee. In mobile optical wireless communication (OWC) networks, however, due to the sensitivity of the received light intensity to the relative displacement between the transmitter and the receiver, the spatio-temporal mobility pattern of UE influences the handover modeling, which in turn complicates the latency management. Moreover, the mobile optical channel is different from the RF counterpart, where the optical channel depends not only on trajectories and fading but also on the spatial relationship between UEs and nontransparent objects. The effect from opaque structures emerges following a time-related human behavior pattern and a space-related pattern of indoor furnishing, as has been highlighted in the previous chapter. In the presence of the wireless channel datasets, a data driven handover mechanism can result in optimal decisions that refrain from frequent handovers and hence overcome the associated latency and reduced throughput limitations. The problem is further complicated, as the OWC channels suffer from abrupt outages in LOS links, and thus, the optical channel gain dataset is highly sparse with huge dynamic range, as elaborated in the previous chapter. Consequently, it is quite hard to learn any handover decision policies.

Hence, in this chapter, we present a data-driven handover policy that smartly determines whether a vertical handover to RF BS is needed or not based on the status of optical links. In order to predict the status of optical channels despite their sparseness, we present novel concepts of channel abstraction, regression, and densification. It should be highlighted that the techniques presented herein can be adopted for integration of other high frequency bands (e.g., mmWave and Terahertz) with the RF band. This is basically since light represents the medium with the shortest wavelength among the aforementioned candidate high frequency bands, and hence, if we can provide a viable channel status prediction and vertical handover framework for RF-optical networks, then other candidate bands, which already present better diffraction capacity than light, can benefit from the same proposed framework.

### 4.1.2 Chapter Organization

The rest of this chapter is organized as follows. First, the system model is described in Sect. 4.2. The handover problem definition is discussed in Sect. 4.3. Based on the generated dataset in Chap. 3, optical link evaluation with outage prediction is presented in Sect. 4.4. In Sect. 4.5, we elaborate on the concept and implementation of the collaborative handover management with prediction of abrupt outages. Finally, simulation results are discussed in Sect. 4.6.

## 4.2  Indoor Layout and Network Model

As shown in Fig. 4.1a, we consider a universal smart office layout model with a $5\,\text{m} \times 5\,\text{m} \times 3\,\text{m}$ room that contains nine desks with a partition height of 1.5, and an entrance centrally located at (2.5 m, 0 m) on the floor, where four light BSs are distributed evenly on the ceiling plane and an RF BS is located on the ground at the corner of the room. Additionally, we consider another meeting room layout as illustrated in Fig. 4.1b, where the room size is also $5\,\text{m} \times 5\,\text{m} \times 3\,\text{m}$ with four light BSs evenly and centrally distributed and one RF BS, with one centrally located meeting desk and eight chairs.

In such a wireless HetNet, let $v$ denote the index of light BS, the single RF BS is indexed by $0$ ($v = 0$) and the $V$ optical BSs are indexed by $\mathcal{V}_{\text{opt}} = \{1, \ldots, v, \ldots, V\}$. The entire BS set is, therefore, $\mathcal{V} = \{0, 1, \ldots, v, \ldots, V\}$. The user is moving in the room in accordance to the mobility model described in Chap. 3. Accordingly, with the mobile channel modeled in Chap. 3, the downlink is supported by VLC-based small BSs and complemented by an RF-based macro BS, while the uplink is enabled by IR and also complemented by RF.

The link assignment is conducted at the RF BS server side, where the decision on handovers is executed at the first time slot in the desired frame index $n$. The duration $I_{\text{H}}$ of handover is counted once the action of handover takes place, which is generally defined by several time frames. The transmitters operate at a constant target rate $R_{\text{T}}$. The BS server conducts link evaluation and handover decision for uplink

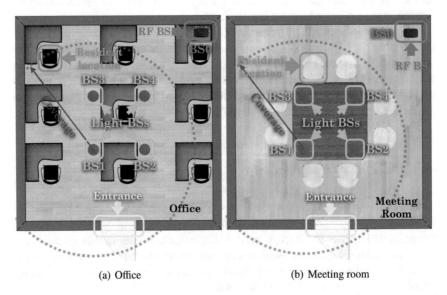

(a) Office                                (b) Meeting room

**Fig. 4.1** Indoor layouts for **a** an office and **b** a meeting room ($5\,\text{m} \times 5\,\text{m} \times 3\,\text{m}$) consisting of four light BSs distributed on the ceiling and one RF BS located at the corner on the floor. All the furnishings and the user body are treated as blockages

and downlink independently since the channel conditions for the two directions are inconsistent. Moreover, we assign the RF channel as an umbrella radio access technology (RAT), therefore, in the link evaluation stage, we only predict the optical channel due to its sensitivity to user's mobility.

### 4.2.1  Achieved Throughput in RF Channels

The channel gain of the RF BS at time $t$ is expressed as

$$H_{\mathrm{RF}}(t) = 10^{-\frac{L\left(d^{(0)}(t)\right)}{10}} |h_{\mathrm{RF}}(t)|^2, \tag{4.1}$$
$$L\left(d^{(0)}(t)\right) = 47.9 + 10\nu \log_{10}(d^{(0)}(t)/d_0) + X \text{ (dB)},$$

where $d^{(0)}(t)$ denotes the distance between the RF BS and the user, the superscript $(\cdot)^{(0)}$ stands for the BS index ($\nu = 0$ for RF BS), and $h_{\mathrm{RF}}(t)$ is the small-scale fading channel modeled by an exponential random variable with mean 2.46 dB [13]; $X$ represents the shadowing component assumed to be a zero mean Gaussian random variable with standard deviation 1.8 dB, $\nu = 1.6$, and $d_0 = 1$ m [13]. The distance $d^{(0)}(t)$ is determined directly from the mobility model described in Chap. 3 based on the relative distance between the user's position and the BS location. The achievable rate from the RF BS indexed by 0 at time $t$ is then expressed as

$$R^{(0)}(t) = W_{\mathrm{RF}} \log_2 \left(1 + \frac{P_{\mathrm{RF}} H_{\mathrm{RF}}(t)}{W_{\mathrm{RF}} \sigma_{\mathrm{RF}}^2}\right), \tag{4.2}$$

where $W_{\mathrm{RF}}$ stands for the bandwidth of RF links, $P_{\mathrm{RF}}(t)$ is allocated power at the RF BS, and $\sigma_{\mathrm{RF}}^2$ is the power of additive white Gaussian noise (AWGN).

### 4.2.2  Achieved Throughput in Optical Channels

The achievable data rate from each optical BS is given by [14]

$$R^{(\nu)}(t) = \frac{W_{\mathrm{opt}}}{2} \log_2 \left(1 + \frac{\left(\eta_{\mathrm{PD}} P_{\mathrm{opt}}^{(\nu)} H_{\mathrm{opt}}^{(\nu)}\right)^2}{\kappa^2 N_0^{(\nu)} W_{\mathrm{opt}} + N_{\mathrm{diff}}^{(\nu)}}\right), \tag{4.3}$$

where $W_{\mathrm{opt}}$ denotes the optical channel bandwidth, $N_0^{(\nu)}$ represents the power spectral density (PSD) of the observed noise, $P_{\mathrm{opt}}^{(\nu)}$ is the transmitted alternating current (AC) optical signal power, $\eta_{\mathrm{PD}}$ denotes the optical to electric conversion efficiency at the receivers, $\kappa$ stands for the conversion between the average electric power of signals and the average optical power, and $\kappa = 3$ guarantees the clipping ratio less than 1% [14]; thus, the clipping noise could be ignored. Also, for simplicity, we assume the noise contribution of optical multi-path effect $N_{\text{diff}}$ due to all the reflective elements can be mitigated through channel equalization [4]. The optical channel gain $H_{\text{opt}}$ accounts only for the line-of-sight (LOS) component, as it contributes the most to the channel's power gain, and it is generated using the framework described in Chap. 3.

## 4.3 Handover Problem Definition

### 4.3.1 Objective Metric

In a mobile optical wireless network, outages due to blockages happen to a single light BS frequently as has been mentioned in Chap. 3. Accordingly, frequent handovers among BSs must be applied to sustain the data stream under QoS guarantee. However, both blockages and handovers cause latency in transmissions. Therefore, the overall network delay becomes a measure on the advisability of outage responses and handover assignment strategies. A wise policy on link assignment encountering frequent outages in movements should skip unnecessary handovers for short-lived outages, and prejudge the optimal link against the upcoming outages, which eventually yields a minimal latency and provides the users with smooth experience.

The actual channel service bit number [7] at time $t$ is expressed as

$$\Gamma_c(t) = \min \left\{ t R_T, R^{(v)}(t) + \Gamma_c(t-1) \right\}, v \in \mathcal{V}. \tag{4.4}$$

Hence, the transmission buffer queue length is

$$Q(t) = t R_T - \Gamma_c(t). \tag{4.5}$$

As the propagation delay is negligible in a small room compared with the network delay, we define the network delay as the duration from $t$ to the time when the actual service reaches $t R_T$, which is the duration of a data package's dwelling in the transmitter's buffer. Therefore, rather than computing a delay, we choose to utilize $Q(t)$ to characterize the network latency.

### 4.3.2 Problem Statement

The representation of QoS refers to guaranteeing a tolerable maximum average queue length $Q_m$ and a target source rate $R_T$. Hence, we formulate the link assignment as Problem 4.1.

**Problem 4.1** (*Minimal Latency*) Given that the objective is to find a strategy $\vartheta^{(v)}(t)$ on link assignment to minimize the overall latency under the QoS constraints in multi-mode RF-optical HetNet, the problem is formulated as

$$\min_{\{\vartheta^{(v)}(t)\}} \quad Q(t) \tag{4.6}$$

$$\text{s.t.} \quad Q(t) \leq Q_m, \tag{4.7}$$

$$\sum_v \vartheta^{(v)}(t) R^{(v)}(t) \geq R_T, \tag{4.8}$$

$$\sum_v \vartheta^{(v)}(t) = 1, \tag{4.9}$$

$$\vartheta^{(v)}(t) = \{0, 1\}, \forall t \in [0, T], \forall v \in \mathcal{V}. \tag{4.10}$$

As aforementioned, this strategy shall minimize the overall network delay by minimizing the queue length at the transmitter's buffer as expressed in (4.6), while satisfying the QoS constraints on the maximum delay represented by the maximum queue length in (4.7) and on a target source data rate $R_T$ in (4.8). At any time slot $t$, only one BS $v \in \mathcal{V}$ will be designated to the user in the multi-mode HetNet as described by (4.9).

### 4.3.3 Overview of Smart Handover Framework

To achieve the objective of *Problem* 4.1, a smart handover framework is illustrated in Fig. 4.2. The presented framework is capable of predicting future status of optical channels, and hence, skip unnecessary handovers. Specifically, handovers can be skipped for short-lived outages that do not significantly accumulate the queue length at the transmitter's buffer. On the other hand, long-lived outages that would significantly increase the queue length triggers a handover decision. The challenge, as will be discussed in Sect. 4.4.2, is that the optical channel for the LOS component is quite sparse, as highlighted in Chap. 3. This fact paralyzes conventional time-series prediction approaches based on deep learning. Hence, special measures need to be taken to overcome such a challenge, as illustrated in Fig. 4.2. First, at the receiver side, the *optical channel gain* is measured and abstracted to obtain *a dense representation of outage events* as will be discussed in Sect. 4.4.3. Then, the *event abstraction sequences* are fed into a *deep-LSTM* (long-short-term-memory) to learn the user's mobility pattern, and hence, predict future outages, according to which, it *evaluates the optical link stability*. This stage is discussed in Sects. 4.4.4 and 4.4.5. Next, the objective (4.6) in *Problem* 4.1 is achieved via *reinforcement learning-based dynamic programming* using the optical link evaluation outcomes, generating the *optimal pol-*

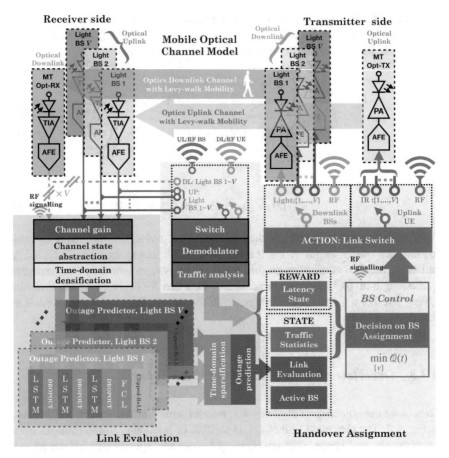

**Fig. 4.2** The structure of the handover framework with opto-electronic front-ends. UE Opt-TX: the uplink optical transmitter of the UE; UE Opt-RX: the downlink optical receiver of the UE, its channel data is signaled through a robust RF uplink to the BS server

*icy on handover assignments*, which comprise the spatio-temporal mobility patterns, as will be detailed in Sect. 4.5. In order to train and test this data-driven framework, the indoor optical channel dataset created in Chap. 3 is utilized.

## 4.4  Channel Event Predictor

A channel event here is an indication of a LOS link outage. More details are given below on the event prediction problem, challenges, and solution approach.

### 4.4.1 Prediction Problem Definition

An intuitive description of this prediction problem is the following: given the normalized channel gain[1] $h^{(v)}(n)$ at frame index $n \in \{1, 2, ..., N\}$ for optical BS with index $v \in \{1, 2, ..., V\}$, the predictor is supposed to forecast the channel gain in the $k-$frame future $h^{(v)}(n + k)$. However, the sparseness of the LOS optical channel gain due to frequent outages, as discussed in Chap. 3, brings trouble to the design of such a predictor since it faces the problem of extremely unbalanced samples and data. This is very different from the RF channel, where the outage probability in fading is far lower than that in the optical counterpart due to the disparity in propagation characteristics. Therefore, unfortunately, we cannot predict the accurate gain of a mobile optical wireless channel. Nevertheless, compared with an accurate regression of the channel gain, a predictor would at least realize certain forecast on occurrence time of outage events. Besides, while making decisions on resource allocation or handover, the accurate forewarning of outages for several upcoming frames matters more if precise channel gain prediction is unrealistic. The prediction of abrupt outages is rather intriguing and realistic for the optical channel given the huge potential of blockages and possibility of being out of sight. So, given $N_E$ events and the input channel gain data $\mathbf{h}$ from all BSs, we formulate the $k-$step prediction problem as

$$\arg\min_{\Theta} \sum_{a=1}^{N_E} \|\hat{T}_a(\mathbf{h}_a | \Theta) - T_a\|, \tag{4.11}$$

where $\hat{T}_a$ is the predicted occurrence time of event $a$ given the input $\mathbf{h}_a$ and model parameters $\Theta$ and $T_a$ stands for the ground true occurrence time. This predictor is supposed to be implemented at the BS side, where the downlink channel information is signaled to the BSs via a stable RF uplink.

Next, we will first discuss how the sparsity of the optical channel affects the predictor's performance during the training stage. Then, we present an approach to abstract the abrupt events and optimize the training sequences from the dataset we created in Chap. 3. Finally, we will describe the complete design of the predictor.

### 4.4.2 Event Prediction Under Sparsity of Channel Gain

The sparsity of the channel gain due to frequent and long-duration outages paralyzes a conventional gradient descent-based optimization to learn the predictor parameters, as it induces several local minima. Regardless of which form we attribute the prediction problem to, the loss function in the training process is associated with the $\ell_2$-norm $\|\mathbf{y} - \hat{\mathbf{y}}\|$ of the ground true value of the normalized channel gain representa-

---

[1]$h^{(v)}(n)$ is the normalized version of $H_{\text{opt}}^{(v)}$ used in (4.3) at frame index $n$. For notation simplicity, and since we discuss prediction only for optical channels, the subscript opt has been omitted.

tion $\mathbf{y}$ and the prediction outcome $\hat{\mathbf{y}}$. However, there are some special points, around which, the loss gradient is very likely to be trapped during the gradient descent-based training. This is a serious issue because we are using the historical records as input to predict a future sequence.

The nature of trapping in a local minimum is that the training loss becomes a constant value, which is independent of predictor parameter $\Theta$.

We first introduce the Type I local minima when the predictor is driven by the parameter $\Theta_0$, as illustrated in Fig. 4.3a. According to the channel statistics presented in Chap. 3, given $N$ samples, the sparsity ratio satisfies $\|\mathbf{y}\|_0/N < 0.5$. Considering a circumstance where under the parameter $\Theta_0$, $\hat{\mathbf{y}}_{\Theta_0}$ is just a $k$-slot-delayed version of the ground true sequence $\mathbf{y}_k$, as the red and blue lines shown in Fig. 4.3b. In this case, the training loss will be not related to the delay $k$ and the corresponding $\Theta$, which means that the gradient of a specific objective function is not sensitive to the error in delay under different parameters. Under this circumstance, the loss is only related to the sparsity of the label, so the training loss has a high probability of being trapped in such local minimum. The prediction outcomes through this training are useless, since the vital indications on the occurrence time of outages cannot be given, as illustrated in red and blue lines in Fig. 4.3b. In this illustration, under the parameters of both $\Theta_0$ and $\Theta_0'$, their losses $J_{\ell_2}$ are the same as shown in Fig. 4.3a. More formally, these Type I local minima are interpreted by the following proposition.

**Proposition 4.1** (Type I Local Minima) *During the gradient descent training, the $\ell_2$-norm-based loss function $J_{\ell_2}(\Theta) = \|\mathbf{y} - \hat{\mathbf{y}}_{\Theta_0}\|_2$ is considered; then there exists $\epsilon > 0$ such that for $\|\Theta - \Theta_0\| < \epsilon$, it holds that*

$$J_{\ell_2}(\Theta) \geq J_{\ell_2}(\Theta_0) = \sqrt{2\|\mathbf{y}\|_1}, \tag{4.12}$$

*where each time slot contributes one proportion of error, and each maximum error is 1 due to normalization. Also, in a binary state channel (i.e., outage or no outage), we have $\sqrt{2\|\mathbf{y}\|_1} = \sqrt{2\|\mathbf{y}\|_0}$ due to $\max_k \|\mathbf{y} - \mathbf{y}_k\|_1 = 2\|\mathbf{y}\|_0$. Hence, after training through several epochs, given the sparsity ratio $\|\mathbf{y}\|_0/N < 0.5$, the $\ell_2$-norm of the prediction loss is trapped around $\sqrt{2\|\mathbf{y}\|_0}$.*

**Proof** The training loss becomes a constant value and is independent of prediction parameter $\Theta$, as it merely depends on the label sparsity. This condition can be expressed as

$$\nabla_\Theta J_{\ell_2}(\Theta_0) = \nabla_\Theta \sqrt{2\|\mathbf{y}\|_1} = 0, \tag{4.13}$$

Therefore the training process is trapped.                                              □

Obviously, the sparser the label, the higher the risk of reaching Type I local minima. We can approximately estimate the chance of being trapped in Type I local minima $J_{\ell_2}(\Theta_0)$ as follow.

**Lemma 4.1** (Approximate risk of Type I local minima) *Given a model parameter similar to $\Theta_0$, the risk of reaching Type I local minima when predicting $k$ time slots forward is approximately*

$$\Pr(k \geq \|\mathbf{y}\|_0) = F_\Delta(k) = \int_0^k f_\Delta(\vartheta)d\vartheta, \tag{4.14}$$

*where $F_\Delta(k)$ denotes the CDF of the signal duration distribution, $f_\Delta$ the PDF of the signal duration distribution, and $\Delta$ the variable duration of signals. This means the probability that the sparsity $\|\mathbf{y}\|_0$ is equal to the signal duration is not longer than the prediction interval.*

**Proof** We can assume that the variable length of each sample sequence is two times of a continuous signal duration; for example, as shown in Fig. 4.3b, we select the sequence from 0 s to 10 s. So, the continuous signal duration in each label sequence will be equal to the $\ell_0$-norm of the label, and the PDF of $\|\mathbf{y}\|_0$ becomes the PDF of the continuous signal duration $f_\Delta(\|\mathbf{y}\|_0)$. Since $\Theta_0$ indicates $\hat{\mathbf{y}}_{\Theta_0} = \mathbf{y}_k$, if the signal duration is shorter that $k$, its loss $J(\Theta)$ will be constant in accordance with (4.12). □

Additionally, the statistics of the channel dataset created in Chap. 3 concludes that about 40% of the signals last shorter than 1 seconds, which negatively impacts the gradient during the training stage. This is a fatal problem for predicting a sudden incident.

On the other hand, we will more likely get a zero sequence, for example the green line in Fig. 4.3b, since the corresponding error gradient might vanish or be trapped in another type of local minima under the parameter $\Theta_1$ shown in Fig. 4.3a.

**Proposition 4.2** (Type II Local minima) *There exists $\epsilon > 0$ such that $\|\Theta - \Theta_1\| < \epsilon$*

$$J_{\ell_2}(\Theta) \geq J_{\ell_2}(\Theta_1) = \sqrt{\|\mathbf{y}\|_1}, \tag{4.15}$$

*due to $\|\mathbf{y} - \mathbf{0}\|_2 = \|\mathbf{y}\|_1$. Also, in a binary state channel: $\sqrt{\|\mathbf{y}\|_1} = \sqrt{\|\mathbf{y}\|_0}$.*

**Proof** In this case, the training loss no longer depends on prediction parameters, since it becomes a constant value $\sqrt{\|\mathbf{y}\|_1}$. Then the gradient gets vanished due to

$$\nabla_\Theta J_{\ell_2}(\Theta) = \nabla_\Theta \sqrt{\|\mathbf{y}\|_1} = 0, \tag{4.16}$$

which means it only depends on the label sparsity. Hence, the training process is trapped here. □

Note that this type of minima is not related to the prediction interval $k$. This situation may take place even as the predictor parameters are initialized. Therefore, we cannot offer any useful estimation on the risk of Type II minima.

Therefore, subject to such a sparsity, the conventional methods for prediction shall not provide satisfactory results when it comes to mobile optical wireless channels.

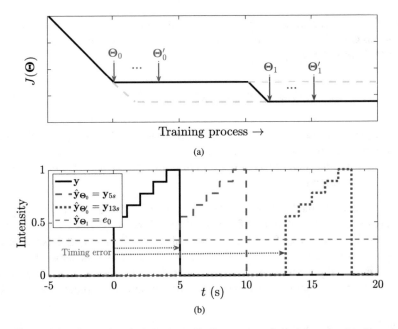

**Fig. 4.3** Demonstration of the channel gain prediction outcomes $\hat{\mathbf{y}}_\Theta$ using conventional regression methods. $\mathbf{y}$ is the ground true value. $\hat{\mathbf{y}}_{\Theta_0}$ and $\hat{\mathbf{y}}_{\Theta_1}$ are the outcomes due to two types of local minima in $J_{\ell_2}(\Theta)$ [15]

### 4.4.3    Event Abstraction and Data Densification

Our goal is to stick with the status quo of the loss functions from existing training methods and to strive to use them directly by transforming the problem in hand. Therefore, we must correlate the loss function and the timing error of the prediction so as to keep the gradient from vanishing under the previously mentioned conditions.

The occurrence of sudden outage in LOS is hidden in the original sparse data. Whenever the channel gain is zeroed or recovered from other states, it is an indication of an abrupt event. Since our concern is the accurate time of occurrence rather than the specific gain, all we have to do is abstracting the intensity data into time-dependent signals. As for the input channel data, we convert it into a timer between every two events as illustrated in Fig. 4.4a. Given a channel gain threshold $h_{\text{th}}$ for outage judgment, we quantify the raw data into binary-state sequences $\mathbf{x}$. Then, as described by (4.17), for the signal duration, the timing indicator $x(n)$ is accumulated positively as time increases; while for the outage duration, the indicator $x(n)$ is accumulated negatively as time increases. Therefore, as for the $n-$th ($n \in [0, N_t]$) sample in a trajectory with $N_t$ samples at BS $v$, the channel event abstraction is generated by

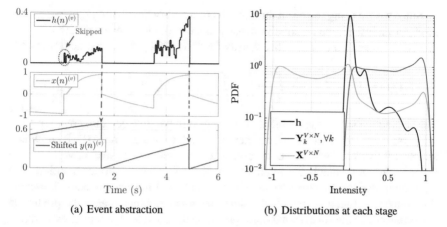

(a) Event abstraction  (b) Distributions at each stage

**Fig. 4.4** **a** The abstraction and time-domain densification of normalized channel gain data for one of the BSs, where $x^{(v)}$ is the abstraction sequence of the BS $v$ after PAR mitigation, and will be treated as one component of the input data, and $y$ stand for the labels of outages. The shown labels have been shifted to $k-$step backward for illustration. Note that the narrow burst signal duration before the first recovery event edge is skipped in the label sequences to gain the potential of bypassing frequent handovers. We used ten iterations of tanh process for all the input and label data. **b** PDF of the sequence in each process in accordance with **a**

$$x(n) = \begin{cases} x(n-1) - 1, & \text{for } h(n) \leq h_{\text{th}} \\ x(n-1) + 1, & \text{for } h(n) > h_{\text{th}} \\ 0, & \text{for } (h(n) - h_{\text{th}})(h(n-1) - h_{\text{th}}) \leq 0, \end{cases} \tag{4.17}$$

where we define $x(0) = 0$ as the initial abstraction unit. The condition $(h(n) - h_{\text{th}}) \cdot (h(n-1) - h_{\text{th}}) \leq 0$ implies an event happens when there is outage at frame $n-1$ and gets recovered at frame $n$, or there is no outage at frame $n-1$ but becomes outage at frame $n$. At the end of each event duration, the absolute value of timing indicator is exactly the corresponding event duration. The label data **y** only counts the duration of the relative event incrementally. For predicting $k-$step forward at the $n-$th sample in a trajectory at BS $v$, the label of outage event abstraction is formulated as

$$y(n) = \begin{cases} y(n-1) + 1, & \text{for } h(n+k-1) \leq h_{\text{th}} \text{ and } h(n+k) \leq h_{\text{th}} \\ 0, & \text{for } h(n+k-1) > h_{\text{th}} \text{ and } h(n+k) \leq h_{\text{th}}, \end{cases} \tag{4.18}$$

where we define $y(0) = 0$. This conversion eliminates the aforementioned local minima, since even under extreme sparsity conditions, the prediction loss in the gradient function always relates to the timing error $k$ as $\|\mathbf{y} - \hat{\mathbf{y}}_k\|_2 = \sqrt{N}k$. This process of abstraction makes the regression of event occurrence time more realistic. In other words, if the triangular waveform $y$ from the event occurrence timer could be regressed, we can obtain forewarning of LOS link absence through the falling edges in the waveform $y$.

Since we are using the event duration, its probability distribution requires a more in-depth analysis. If the cumulative density function (CDF) of event duration $F_\Delta$ and its derivative exist, the probability density function (PDF) of the abstracted data $y$ is

$$f_Y(y) = (F_Y(y))' = \frac{d}{dy} \int_0^y \frac{1}{y-x} \int_x^y f_\Delta(v) dv dy = \frac{1}{y}(F_\Delta(y) - F_\Delta(0)).$$
(4.19)

The abstraction solves the problem of sparseness in the time domain; however, from the perspective of the timer values, there is a peak-to-average ratio (PAR) problem, which still leads to unbalanced samples since the values of $x_n$ and $y_n$ will be large for long-duration-signals but relatively small for those with shorter duration. The regression loss for the small triangle components will not occupy enough proportion in the loss function since their values are smaller, which makes the prediction performance for the short signal duration worse than that for the longer duration.

Therefore, we need further PAR mitigation on the abstracted occurrence timer. By employing the hyperbolic tangent (tanh) upon the triangular waveforms of $\mathbf{x}$, $\mathbf{y}$ as

$$x(n) \leftarrow \tanh\left(x(n)/x_{\text{up}}\right), \text{ for } x(n) > 0,$$
$$x(n) \leftarrow \tanh\left(x(n)/x_{\text{low}}\right), \text{ for } x(n) < 0, \qquad (4.20)$$
$$y(n) \leftarrow \tanh\left(y(n)/y_{\text{up}}\right),$$

where $x_{\text{up}}$, $y_{\text{up}}$, and $x_{\text{low}}$ stand for estimates of the maximum and minimum values of the sequences before applying this compression. Only compressing these sequences for one time may not reduce enough PAR if the duration varies in a large range. The compressed sequences can be compressed again. Thus, the PAR mitigation should be iterated for appropriate number of times until the PDF of input data and labels get fairly flattened as shown in Fig. 4.4b. The corresponding PDF after deploying tanh once is given by

$$f_T(x) = \frac{F_\Delta(\tanh^{-1}(x)) - F_\Delta(0)}{(1-x^2)\tanh^{-1}(x)}. \qquad (4.21)$$

Accordingly, we abstract the abrupt event as the input data $\mathbf{X}^{V \times N}$ of $N$ samples and the label data $\mathbf{Y}_k^{V \times N}$. We use the abstraction of the real-time sample of the channel as input, no matter how many intervals to predict.

### 4.4.4 Event Regression

After procedures of time-domain abstraction and densification, we obtain the input data $x^{(v)}(n)$ at $n$th sample for LSTM; then for each BS, the prediction is conducted independently but with the same input data $\mathbf{z}^\top(n)$ of size $V + 1$ and combined with abstractions $\mathbf{x}^\top(n)$ of all the BSs and an auxiliary tracing timer as $t_n = \tanh(t/T)$, where $t \in \mathbb{N}^+$ is the time of the first time slot inside each frame index in an entire trajectory, and $T$ is a scale factor denoted by the mean duration of mobility. The auxiliary tracing timer $t_n$ is very useful for learning the return tendency in the macro pattern, since in different samples, the channel status changes similarly during the same trajectory stage.

To limit the time cost in the regression stage, each BS carry out the prediction independently so that the requirement in both neural network depth and size will be reduced. However, each single BS still needs the event abstraction from all of the BSs since the data from any single BS cannot represent the whole picture. For instance, when the user encounters a corner and starts to turn the UE orientation, there will be a handover from the current BS to its next BS on the new direction. What the current BS will observe is just an outage, but cannot capture which BS is the next one and where the user has turned to. Hence, the data from all BSs makes learning the mobility pattern from the channel gain data under an indoor structure possible for the predictor. Therefore, the final input data will be $\mathbf{Z}^{(V+1)\times N}$ for every BS, and the label for the BS $v$ will be $(\mathbf{y}_k^{(v)})^{1\times N}$.

As for learning the long-term dependency among all BSs, we set up several LSTM layers with dropout and one fully connected layer (FCL) and one clipped rectified linear unit layer (ReLU) in what concerns the constraints on the output magnitude. We adopt the LSTM in accordance with the universal structures [16], where in the $p$-th layer and BS $v$, $\mathbf{W}_i^{p,v}, \mathbf{W}_f^{p,v}, \mathbf{W}_o^{p,v}, \mathbf{W}_c^{p,v}$ stand for the weight matrices of the input vectors for the input gate, forget gate, output gate, and the cell with number of LSTM units equal to $C$, respectively. The weights in the first layer belong to $\mathbb{R}^{C\times V+1}$, and to $\mathbb{R}^{C\times C}$ for the other layers; $\mathbf{U}_*^{p,v} \in \mathbb{R}^{C\times C}$ represents the matrices of weights regarding their hidden states; $\mathbf{b}_*^{p,v} \in \mathbb{R}^C$ denotes their biases; and $\mathbf{V}^l, \mathbf{b}_f^l \in \mathbb{R}^C$ represent the weight and bias vectors for FCL.

Given a training set with $N$ samples, our objective is to minimize the following function:

$$J^{(v)}\left(\mathbf{\Theta}^{(v)}\right) = \frac{1}{N}\|\hat{\mathbf{y}}^{(v)}(\mathbf{Z}|\mathbf{\Theta}^{(v)}) - \mathbf{y}^{(v)}\|_2^2 + \frac{\lambda}{2}\|\mathbf{\Theta}^{(v)}\|_2^2, \qquad (4.22)$$

where $\mathbf{\Theta} = \{\mathbf{W}_*, \mathbf{U}_*, \mathbf{V}_*, \mathbf{b}_*\}$ denotes all the parameters to be estimated and $\lambda$ is the $\ell_2$-regularization factor. We also adopt Adam [17] for first-order gradient-based optimization of the objective function (4.22) in Algorithm 5, with the initial learning rate $\eta = 10^{-3}$, gradient decay factor $\xi_1 = 0.9$, squared gradient decay factor $\xi_2 = 0.999$, $\varepsilon = 10^{-8}$, mini-batch size $S = 128$, and the number of epochs $Q = 500$.

In Algorithm 5, from line 11 to line 15, we update the biased first moment estimate $\mathrm{m}_t$ at time step $t$ and the biased second raw moment estimate $\mathrm{v}_t$, then compute the bias-corrected first moment estimate $\hat{\mathrm{m}}_t$ and the bias-corrected second raw moment estimate $\hat{\mathrm{v}}_t$, and finally, update $\mathbf{\Theta}^{(v)}$.

---

**Algorithm** 5. Training for Prediction of Event Abstraction at BS $v$

---

**Data:** $\mathbf{Z}^{(V+1) \times N}$, $(\mathbf{y}_k^{(v)})^{1 \times N}$

**Result:** Optimal parameters $\Theta^{(v)}$

1: **Initialization:** $\Theta^{(v)}$, where $\mathbf{W}_i^{*,v}$ are initialized with Glorot [18]; $\mathbf{W}_f^{*,v}, \mathbf{W}_o^{*,v}, \mathbf{W}_c^{*,v}$ are initialized with the orthogonal matrix [19]. Iteration index $q = 1$. Time step $\mathbf{t} = 0$. The first moment vector $\mathbf{m}_0 = 0$. The second moment vector $\mathbf{v}_0 = 0$.

2: **while** $q \neq Q$ **do**

3:     Initialize: $s = 1$

4:     **while** $s \neq S$ **do**

5:         **for** each samples $\mathbf{Z}$ and $\mathbf{y}_k^{(v)}$ in mini-batch $s$ **do**

6:             **Feed Forward**: Get $\hat{\mathbf{y}}^{(v)}(\mathbf{Z}|\Theta^{(v)})$

7:             **Back Propagation**: Compute $\nabla_\Theta J^{(v)}(\Theta^{(v)}|s)$

8:         **end for**

9:         **Weight and bias update:**

10:            $\mathbf{t} = \mathbf{t} + 1$

11:            $\mathbf{m}_t \leftarrow \xi_1 \mathbf{m}_{t-1} + (1 - \xi_1) \sum_s \nabla_\Theta J^{(v)}(\Theta^{(v)}|s)$

12:            $\mathbf{v}_t \leftarrow \xi_2 \mathbf{v}_{t-1} + (1 - \xi_2) \sum_s \nabla_\Theta J^{(v)}(\Theta^{(v)}|s)$

13:            $\hat{\mathbf{m}}_t \leftarrow \mathbf{m}_t / (1 - \xi_1^t)$

14:            $\hat{\mathbf{v}}_t \leftarrow \mathbf{v}_t / (1 - \xi_2^t)$

15:            $\Theta^{(v)} \leftarrow \Theta^{(v)} - \eta \hat{\mathbf{m}}_t / (\sqrt{\hat{\mathbf{v}}_t} + \varepsilon)$

16:     **end while**

17: **end while**

---

### 4.4.5  Event Sparsification

After regression, the predicted event abstraction sequences will be sparsified back to their original form of scattered indications of outages. This can be carried out as follows. Given the definition in (4.18), each falling edge in the event abstraction waveform represents an event, so the sparsification stage treats each falling edge that varies numerically more than a threshold $y_{th}$, designated by the minimal event interval, as an event indication following

$$\hat{\mathbf{i}}(n) = \begin{cases} 1, & \text{if } y(n) - y(n-1) > y_{th} \\ 0, & \text{otherwise.} \end{cases} \tag{4.23}$$

The sequences $\hat{\mathbf{i}}(n) \in \{0, 1\}$ represent outage events, which are Boolean indices of abrupt events that are obtained at this stage. Assuming the ground true sample of indications as $\mathbf{i}(n)$, $y_{th}$ should be decided as

$$\underset{\{y_{th}\}}{\arg\min} \frac{1}{N} \sum_{n=1}^N \left| \mathbf{i}(n)(y(n)|y_{th}) - \hat{\mathbf{i}}(n)(y(n)|y_{th}) \right|. \tag{4.24}$$

In fact, if the regression of the waveform is accurate in shape, the falling edge will be easy to recognize. While a more advanced algorithm using the convolutional neural network (CNN)-based approach for such recognition could improve the prediction accuracy, the extra time cost cannot be ignored since the prediction is conducted in real-time. Given a certain timing delay error for a predicted event, the total prediction time cost must be less than the prediction interval; and the extra time cost throughout the sparsification process is treated as an additional timing delay error. Therefore, we would rather choose a simple way in this stage if the core regression accuracy is adequate.

## 4.5 QoS-Guaranteed Handover Assignment

The strategies on handovers for minimizing the overall latency is generated by Q-learning algorithm [20] with outage prediction using the channel dataset created in Chap. 3. The handover decision is also conducted at the BS server side like the prediction. The states, actions, and rewards are defined from the perspective of the BS.

### 4.5.1 Definition of States

#### 4.5.1.1 State of Link Evaluation

First, the outage prediction is used to produce the link evaluation $s_b$ as one part of the states. The sequence $\text{o}^{(v)}(n) \in \{0, 1\}$ stands for the forewarning of outages for optical AP $v$ at frame $n$. Once an active falling edge in the regressive outcomes is recognized where $\text{i}^{(v)}(n) = 1$, the forewarning of optical outage in this link will be held as '0' until the actual outage occurs. Based on $\text{o}^{(v)}(n)$, the link evaluation yields an optimal BS assignment as $s_b \in \mathcal{V}$, which contains all the AP indices; $s_b$ is defined as follows

$$s_b = \begin{cases} 0, & \text{for } h^{(v)}(n) \leq h_{\text{th}}, \forall v \in \mathcal{V}_{\text{opt}} \\ \arg\max_{\{v\}} h^{(v)}(n)\text{o}^{(v)}(n), & \text{otherwise.} \end{cases} \tag{4.25}$$

If all the optical BSs suffer from outages, it yields $s_b = 0$ indicating that only the RF channel is stable; otherwise, $s_b$ implies the optical BS with the highest channel gain and without future outage risks. This is a passive state since no action can nudge any change of the channel gain, which depends on the user's mobility and the indoor environment.

#### 4.5.1.2 State of Active BS

Second, we define the active BS index $s_a$ as representing the assigned BS at each frame; thus, $s_a \in \mathcal{S}_a = \{0, 1, 2 \ldots, V\}$, a set of size $V + 1$.

### 4.5.1.3    State of Data Traffic

The last part is the traffic state $s_t$ that quantifies the actual data transmission progress in the space $S_t = \{\Gamma_1, \ldots, \Gamma_m, \ldots, \Gamma_M\}$ of size $M$.

Defining the expectancy of the total source traffic in a trajectory with the average duration $T_E$ as $\Gamma_E$

$$\Gamma_E = \Gamma_c(T_E), \tag{4.26}$$

with $\Gamma_c(t)$ introduced in (4.4) and $T_E$ can be calculated as

$$T_E = \frac{t_{s,min}^\alpha}{1 - \left(t_{s,min}/t_{s,max}\right)^\alpha} \cdot \left(\frac{\alpha}{\alpha - 1}\right) \cdot \left(\frac{1}{t_{s,min}^{\alpha-1}} - \frac{1}{t_{s,max}^{\alpha-1}}\right). \tag{4.27}$$

At the first time slot of frame $n$, we have the traffic $\Gamma_c(n)$, and we define $s_t$ as

$$s_t = \begin{cases} \Gamma_m, & \text{for } \Gamma_c(n) \in \left[\frac{(m-1)}{M}\Gamma_E, \frac{m}{M}\Gamma_E\right) \\ \Gamma_M, & \text{for } \Gamma_c(n) > \Gamma_E, \end{cases} \tag{4.28}$$

which means $\Gamma_c(n)$ will be categorized to state $\Gamma_m$, if $\Gamma_c(n) \in [\frac{(m-1)}{M}\Gamma_E, \frac{m}{M}\Gamma_E)$; and particularly for $\Gamma_c(n) > \Gamma_E$, the corresponding $\Gamma_m$ is limited by $\Gamma_M$. This state component is important, since it encodes the transmission progress as well as the channel condition over time. For instance, assuming an ideal channel without outages during the entire trajectory, its $s_t$ should switch monotonously from $\Gamma_1$ at the first frame to $\Gamma_M$ at the last frame; however, if it suffered from the considerable duration of outages in a BS, its $s_t$ would stay in a relatively low state.

### 4.5.1.4    State Space

So, we conclude that a state $s$ belongs to the space $S = S_b \times S_a \times S_t$, whose dimensionality is $(V + 1)^2 M$.

## 4.5.2   Definition of Actions

We define an action $a$ based on the action space $\mathcal{A} = \mathcal{V} = \{0, 1, 2, \ldots, V\}$, which features the index of the designated BS. If the actions are different in two adjacent frames $a(n) \neq a(n - 1)$, it will incur the handover process starting from the current frame denoted by $n_H = n$ to the end of this handover process. Given handover overhead as $I_H$, the action will not get changed during the handover process as $a(n + 1), \ldots, a(n_H + I_H) = a(n)$; otherwise, the action will be chosen following the $\epsilon$-greedy algorithm with the coefficient $\epsilon_Q$. Meanwhile, during the handover process, the data transmission is suspended, so we can re-write (4.2) and (4.3) at the first time slot in frame $n$ as

$$
R^{(v)}(n) = \begin{cases} \mathbb{H}(n) W_{\mathrm{RF}} \log_2 \left( 1 + \dfrac{P_{\mathrm{RF}} H_{RF}(t)}{W_{\mathrm{RF}} \sigma_{\mathrm{RF}}^2} \right), & \text{for } v = 0, \\[4mm] \mathbb{H}(n) \dfrac{W_{\mathrm{opt}}}{2} \log_2 \left( 1 + \dfrac{\left( \eta_{\mathrm{PD}} P_{\mathrm{opt}}^{(v)} H_{\mathrm{opt}}^{(v)} \right)^2}{\kappa^2 N_0^{(v)} W_{\mathrm{opt}} + N_{\mathrm{diff}}^{(v)}} \right), & \text{for } v \in \mathcal{V}_{\mathrm{opt}}, \end{cases}
\tag{4.29}
$$

where

$$
\mathbb{H}(n) = \begin{cases} 0, & \text{for } n \in [n_{\mathbb{H}}, n_{\mathbb{H}} + I_{\mathbb{H}}] \\ 1, & \text{otherwise.} \end{cases}
\tag{4.30}
$$

However, no action could affect the actual UE position or motion, since no matter what action is taken, the current $s_b$ will transfer to another state following a transient probability in compliance with the mobility patterns, which, however, are entirely independent of the action at the BS side.

### 4.5.3 Definition of Rewards

Given the objective of minimizing the queue length in (4.6) so as to mitigate the overall latency, we define the reward imposed on the differential queue length, which means the queue increment or reduction between two adjacent frames in the transmitter's buffer. Specifically, the reward at frame $n$ is defined according to four states of the differential queue length

$$
r(n) = \begin{cases} r^{(a)}, & Q(n-1) = Q(n) = 0 \\ r^{(b)}, & Q(n-1) > Q(n) \\ r^{(c)}, & Q(n-1) = Q(n) \neq 0 \\ r^{(d)}, & Q(n-1) < Q(n) \end{cases}
\tag{4.31}
$$

where $Q(n)$ denotes the queue length at the first time slot in this frame, $r^{(a)} > r^{(b)} \in \mathbb{Z}^+$ are positive rewards, while $r^{(c)} > r^{(d)} \in \mathbb{Z}^-$ perform penalties. The reduction in buffer queue and the maintenance of an empty buffer result in rewards; otherwise, it turns into penalties.

The reward definition in (4.31) enables a strategy that focuses on eliminating the buffer queue. The result of its accumulation evaluates the strategy on maintaining the smallest buffer queue against different transmission processes and channel predictions. Since keeping an empty buffer queue results in the minimal network latency, the highest reward is given to keeping such status. When there are short burst outages in an optical channel, some handovers can be skipped but the reward remains to be high since the buffer queue will not be augmented too much and can be reduced quickly, and also the transmission interruption due to handover process is bypassed. For vertical handovers, when all optical BSs are predicted to meet long-outages in the near future, it should choose an RF BS to ensure that the queue length does not get accumulated too fast, even if the optical BSs could still provide high data rate temporarily; when some of the optical BSs recover from outages, it should be motivated to get back to an optical BS to reduce the buffer queue at the fastest speed.

For multi-mode connections, it is unsuited to define rewards directly based on the data rate, which fluctuates dramatically in presence of frequent outages. Under the trajectories with more outages, the reward from data rate will be accumulated slowly; while under the trajectories with better conditions, the reward from data rate will be very high for all actions. Such definition might work if there is not too much diversity in the training set. However, it makes the convergence of policy very hard if the channel conditions in the training episodes have large variance, which is inherited from the nature of human mobility.

### 4.5.4   Q-learning-based Handover Policy

Finally, we can deploy the Q-learning algorithm to obtain a policy $\pi(s, a)$ that maximizes the overall reward as

$$\arg\max_{a \in \mathcal{A}} Q\left(\pi(s, a)\right). \tag{4.32}$$

The learning algorithm is detailed in Algorithm 6, where we update the knowledge $Q$ as:

$$Q(s, a) \leftarrow \begin{cases} Q(s, a) + \alpha \left(r + \gamma \max_{a'} Q\left(s', a'\right) - Q(s, a)\right), \\ \qquad \text{for } \mathbb{H}(n) = 1 \text{ or } n = n_{\mathbb{H}} \\ Q(s, a), \ \text{for } n \in [n_{\mathbb{H}} + 1, n_{\mathbb{H}} + I_{\mathbb{H}}]. \end{cases} \tag{4.33}$$

where $\alpha$ and $\gamma$ stand for the initial learning rate and discount factor, respectively. In Algorithm 6, we need an environment emulator that executes the entire network simulation. Also, we define a complete trajectory as an episode, in which the UE starts at the entrance and ends up at the exit. However, different from the conventional Q-learning [20], we do not update the Q-table during the period starting from the second frame of handover process until it is done. Although the transmission is suspended during the handover process, the prediction state $s_b$ will still get changed; if $Q(s, a)$ gets updated for the whole process, the $Q(s, a)$ associated with the prediction state $s_b$ during this process will decrease rapidly due to the increment in queue length, which causes an incorrect correspondence of the channel prediction state with the $Q$ value.

---

**Algorithm** 6. QoS-Guaranteed Handovers

---

**Require:**
1:  $E$: Environment emulator;
2:  $\mathcal{A}$: action space;
3:  $S$: state space;
4:  Outage Prediction.
**Ensure:** $\arg\max_{a\in\mathcal{A}} Q\left(\pi(s,a)\right)$
5:  Initialize $Q(s,a)$ arbitrarily
6:  Initialize indoor layout
7:  **for** each episode **do**
8:      Reset mobility generator
9:      Initialize state $(s_b, s_a, s_t) = (0, 0, \Gamma_1)$
10:     **for** each frame $n$, and the user remains indoor **do**
11:         **if** in handover process **then**
12:             Keep $a(n) = a(n-1)$
13:         **else**
14:             Choose $a_n$ following $\epsilon-$ greedy $\pi^{\epsilon Q}(s,a)$
15:         **end if**
16:         Execute $a_n$ and observe $s_a'$, $s_q'$ and $r$
17:         Execute outage prediction, yielding $s_b'$
18:         $s' \leftarrow (s_b', s_a', s_t')$
19:         **if** not in handover process **or** at the first frame of handover process **then**
20:             Update $Q(s,a) \leftarrow Q(s,a) + \alpha(r + \gamma \times \max_{a'} Q(s',a') - Q(s,a))$
21:         **end if**
22:         $s \leftarrow s'$
23:         Renew the trajectory for frame $n+1$
24:     **end for**
25: **end for**
26: **return** $Q$

---

# 4.6  Numerical Results and Discussions

## 4.6.1  *Parameter Setup*

The mobility model and its parameters defined in Chap. 3 are used here. When training with a real trajectory, we need to focus on the moving periods, but keeping the integrity of the mobility model is also essential. So in order not to increase the amount of data excessively while ensuring the integrity of the model, the sampling rate during stationary states was reduced 1,000 times.

For the optical networks, we set the bandwidth of the optical channel to $W_{\mathrm{opt}} = 40$ MHz, and assume the noise density of optical current as $N_0 = 10^{-21}$ A$^2$/Hz for simplicity; the optics-electronic efficiency is set up to $\eta_{\mathrm{PD}} = 0.53$ A/W; and for the RF network, we set the noise density to $-57$ dBm/MHz, and the bandwidth $W_{\mathrm{RF}}$ to 20 MHz. The transmitting signal power of each downlink light BS is 12 W, while the signal power of UE-IR uplink is fixed to 0.5 W. The allocated powers for the umbrella RF are 0.1 W and 0.01 W for downlink and uplink, respectively.

For the outage predictor, we used the dataset with 5,000 traces for training the outage predictors using three identical layers of LSTM, and other 1,000 traces were used for testing the predictor. We also set the number of the LSTM units in a layer $C$ to 64, $\lambda$ to $10^{-4}$, and dropout rate to 0.2. As for Adam optimization, we set the initial learning rate to $\eta_L = 10^{-3}$, gradient decay factor $\xi_1 = 0.9$, squared gradient decay factor $\xi_2 = 0.999$, $\epsilon_L = 10^{-8}$, mini-batch size $S = 128$, and number of epochs $Q = 500$. All the hyper-parameters are tuned manually based on the principle of employing the minimal size LSTM to provide a proper performance. We let the predictor skip the burst momentary signals for the label, as shown in Fig.4.4a.

As for the parameters in Q-learning, we set $M = 4$, $\alpha = 0.3$, $\gamma = 0.9$, $\{r^{(a)}, r^{(b)}, r^{(c)}, r^{(d)}\} = \{3, 2, -2, -3\}$, and $\epsilon_Q = 0.1$, which drops exponentially. The duration of the handover process is fixed as five frames (500 ms) [4].

For training and evaluating our introduced data-driven algorithm, we use the office scenario shown in Fig. 4.1a; while we also set up another layout, the meeting room scenario illustrated in Fig. 4.1b, for the sake of examining the applicability of our framework to diverse settings.

We first test the performance of the predictor under different prediction intervals, as shown in Fig. 4.5. According to our intuition, the farther we predict, the worse performance we should get. For intervals ranging from 1 second to 10 seconds, there has been a dramatic change in the middle. Outside of these two boundaries, the testing performances are slightly distinguishable. For example, the mean-squared error (MSE) performances of the intervals of 1 second and 0.1 second are almost the same, such that we can choose either of these two targets; but choosing 1 second means we can gain a longer perspective on the channel dynamics. Therefore, in the spirit of maximizing the exploitation of the regression stage, we choose 1 second as the target of prediction interval. It should be noted that the case where the interval is 100 seconds is not realistic, as our previous statistical analysis indicated that there is only a 5% chance that a continuous signal event will exceed 5 seconds. Therefore, to achieve such a long-term prediction interval, the regression must be forced to learn from the process that crosses several continuous residence and motion periods or even different traces, but such learning is impractical due to the very rare correlation. This also explains why there is a baseline of prediction interval. Also, according to the statistics as mentioned above, choosing 1 second prediction interval satisfies the demand of sample diversity since most events last longer than 1 second. Moreover, even though the goal of 100 second prediction interval is unreasonable, its MSE drops down at the same rolling speed as other reasonable setups during the first several epochs as observed in Fig. 4.5. The cause of this false appearance is consistent with our previous description in Sect. 4.4.2. The farther the event to be predicted is located, the harder it is to link the occurrence of timing error directly with the loss function.

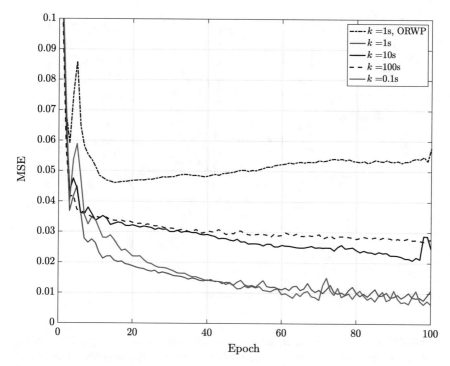

**Fig. 4.5**  MSE losses during the training process

## 4.6.2  *Prediction Interval*

## 4.6.3  *Prediction Performance*

We define the hit rate as the ratio of the counts of valid event hit to the total event number. Supposing that an event occurs at $n_0$, then one valid hit is conditioned on: $\sum_{j=n_0-k/2}^{n_0+k/2} \hat{\mathbb{i}}(j) = 1$, which means that there exists only one event indication within $\pm 0.5k$ steps centered on the ground true occurrence time. Also, all the misjudgments other than the valid hits are regarded as false triggers; the false trigger rate is defined as the ratio of the counts of false triggers to the total event number. In addition, for a valid hit $\hat{\mathbb{i}}(j)$, we define its absolute timing error as: $|j - n_0|$, which indicates the difference between the estimated occurrence time $j$ and the true time $n_0$. Figure 4.6 illustrates these definitions. The overall hit rate of forewarning of abrupt outage events is 91.55%. For the downlink, the mean absolute timing error is 74 ms for outage events. Similarly, for the uplink, the mean absolute timing error is 86 ms for outage events. The cases in which the predictor failed to capture the upcoming events are usually induced by the sudden movements from a sojourn state or by an unexpected blockage from the user body or surroundings under the randomness of orientation. Such randomness in the micro mobility pattern is very hard to exploit.

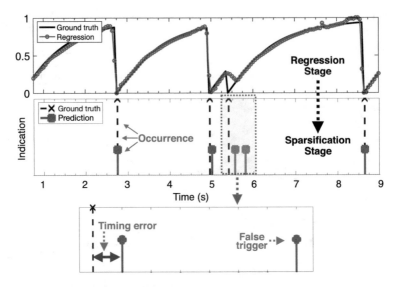

**Fig. 4.6** Examples of prediction performance metrics

Finally, as a comparison, we applied the orientation random waypoint (ORWP) mobility model [21], where the user randomly chooses the step length and destinations from the areas outside the furnishings, and the corresponding trajectory is then used to generate the channel dataset. Under the aforementioned configuration and using same dataset size, the predictor failed to learn any valid information from this dataset. The testing loss of the relevant regression stage is shown in Fig. 4.5a, where the actual regression result under the ORWP-based dataset is almost a straight line, which did not get any better even if the regularization is bypassed. This comparison further verifies the importance of the mobility model selection since the spatio-temporal features are crucial to a learning-based predictor. In fact, what this regression learns is the spatio-temporal progression of the mobility, and furthermore, the relationship with the blockage layout. The auxiliary tracing timer $t_n$ is very important, as it forces the predictor find the relationship between the event abstractions and mobility stage ($T_d$), which corresponds to the macro mobility pattern as mentioned in Chap. 3, and the macro mobility pattern lays the foundation of predictability.

### 4.6.4  Trace Information

The only information used at the BS side is the channel gain received from the photo-diodes. We abstract the pattern in the channel changes due to UE mobility by leveraging the channel gain directly rather than merely considering UE trajectories and orientations. The trajectory data will not be used when training the predictor because the channel gain data is affected by many factors including the UE location, orientation, motion, and furnishings. Therefore, location (GPS) data may mislead the predictor by merely linking locations with channel states. Second, when applying this predictor to real indoor communications, accurate location information is extremely tough to obtain, given the high sensitivity of light intensity to relative displacements and orientations.

### 4.6.5  Handover Assignment

Next, we evaluate the performance of the entire handover framework. Two examples of office rooms for performance evaluation are shown in Figs. 4.7 and 4.8. Any differences between the two office layouts in channel sparseness and outage probability are mainly due to the difference in mobility details. Based on the evaluation dataset, we compared the data-driven handover scheme with the instantaneous handovers in multi-mode optical standalone networks, RF standalone network, and RF-optical HetNets. In this evaluation scenario, we set $R_T = 20$ Mbit/s for uplink and $R_T = 100$ Mbit/s for downlink.

In Figs. 4.8-(b), (c), (e), and (f), the data traffic curves of $\Gamma_c$, representing the served bits under our data-driven handover policy, outperform the instantaneous handover schemes in all types of networks. Please note that the RF standalone network can provide a stable connection; however, the inadequate capacity makes the overall latency rise up over time. The circumstances in the standalone optical networks are different since the instantaneous capacity of each BS while in the FOV exceeded $R_T$, but the frequent outages and handovers still lead to significant delays. Also, the RF-optical HetNet with instantaneous handover does not always perform

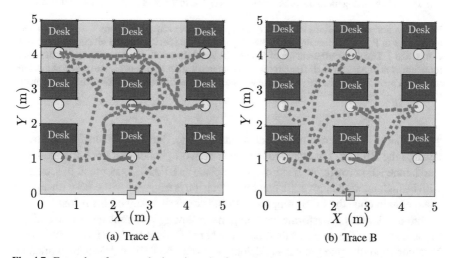

(a) Trace A                                         (b) Trace B

**Fig. 4.7**  Examples of two synthetic trajectories for evaluating network performance

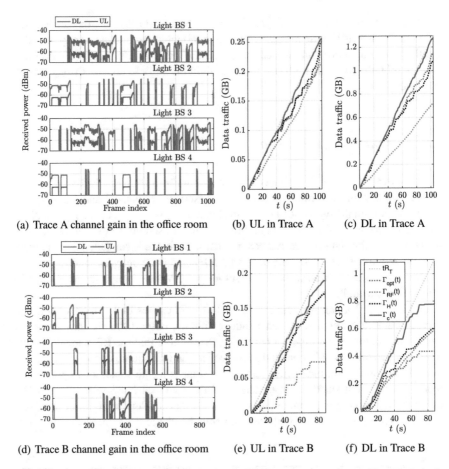

(a) Trace A channel gain in the office room    (b) UL in Trace A    (c) DL in Trace A

(d) Trace B channel gain in the office room    (e) UL in Trace B    (f) DL in Trace B

**Fig. 4.8** **a** and **d** Illustrations of the received optical power sequences along with two synthetic trajectories in Fig. 4.7 for evaluating network performance specially in the office room setup. **b**, **c**, **d**, **e** and **f** The data traffic curves of uplink (UL) and downlink (DL) correspond to the office room traces A in **a** and B in **d**. $t R_T$ expresses the source traffic; $\Gamma_{opt}$ denotes the optical standalone services; $\Gamma_{RF}$ represents the RF standalone services; $\Gamma_H$ is the RF-optical HetNet with instantaneous handovers

better than the standalone ones, such as in Fig. 4.8c, due to the intensive handovers towards the RF link. For standalone optical network, handovers are horizontal. For standalone RF network, no handover takes place as RF LOS is not lost. In scenarios with instantaneous handovers, a handover decision is triggered to the BS with highest channel gain once the LOS component of the currently assigned BS is lost.

Table 4.1 gives a performance comparison among different handover policies implemented on the entire evaluation set, where "Optical SA", "RF SA", and "Het-Net" correspond to the optical standalone network, the RF standalone network, and the RF-optical HetNet with instantaneous handover, respectively; "Only OP" refers

**Table 4.1**  Average network latency on the evaluation sets

| Direction | Optical SA[a] (ms) | RF SA[a] (ms) | HetNet[b] (ms) |
|-----------|--------------------|--------------------|--------------------|
| Uplink | 55.4 | 28.1 | 31.4 |
| Downlink | 52.4 | 49.6 | 42.7 |
| Direction | Only OP[c] (ms) | w/o OP [d] (ms) | w/OP [e] (ms) |
| Uplink | 29.7 | 30.9 | 10.6 |
| Downlink | 44.5 | 39.3 | 13.2 |

[a]Standalone networks with instantaneous handovers
[b]RF-optical HetNets with instantaneous handovers
[c]RF-optical HetNets, using outage prediction only without Q-learning
[d]RF-optical HetNets, using Q-learning only without outage prediction
[e]The complete data-driven handover framework

to handover decision based only on channel prediction without adopting a Q-learning policy, "w/o OP" stands for the RF-optical HetNet with Q-learning-based handover assignment without outage prediction, and "w/ OP" represents the complete handover framework with outage prediction and Q-learning-based handover assignment. When deploying the Q-learning without outage predictions, where $s_b$ merely gives the BS with highest channel gain in optical or RF, the performance is very similar to that of HetNet with instantaneous handover. This means that simple reinforcement learning cannot provide sufficient modeling of the channel variations because it failed to link the latency with the current channel states. Meanwhile, when making handover decisions based on predictions only without Q-learning, the performance is inadequate since it made the UE dwell in the RF BS for most of the time when outages were predicted to be frequent; also the overall network delay due to handovers was not considered.

When it comes to the data-driven handover framework, the $\Gamma_c$ curves in Fig. 4.8b, c almost kept sticking with the source traffic curve all the time except during some handover processes and inevitable outages. Figure 4.9 shows the evolution in the BS assignment strategies on the evaluation set along with the training progress of the Q-learning. One can appreciate how tortuous the process of achieving such performance is, where it started from the totally random choices then gradually mastered leveraging the outage prediction to minimize the latency by skipping unnecessary handovers. Besides, the handover strategies in uplink and downlink are slightly different owing to the inconsistent channel states.

Next, we investigate the performance of the data-driven handover framework under various channel conditions. Intuitively, Trace B illustrated in Fig. 4.8d represents a very extreme scenario of high difficulty to the optical links due to the crucial outage probability. This is reflected in Fig. 4.8e, where the optical links support the least traffic. Again, such little traffic from optical links is caused by the aggravated outages rather than the channel gain. Fortunately, the session indeed gained the traffic at most of the intervals where the optical links can maintain a considerable capacity. Particularly in Fig. 4.8e, the session leveraged for most of time the RF channel, but it decisively switched to optical channels once the link evaluation implied a feasible

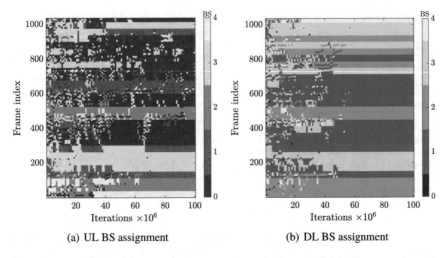

(a) UL BS assignment                    (b) DL BS assignment

**Fig. 4.9**  BS assignment details in the evaluation example Trace A of the office room shown in Fig. 4.8a–c at different Q-learning training stages for uplink and downlink, respectively

duration in optical BSs. However, if the predictor fails to forewarn a long outage duration, as shown in Fig. 4.8f, from 68 s to the end, the BS held the connection of optical channel since the outage prediction missed in the link evaluation ultimately misled the handover decision into dwelling in the optical BSs. Since one part of the state space consists of the prediction outcomes, the prediction accuracy is vital to the overall performance under extreme channel conditions.

In addition to predictions, the definition of reward is determinant when dealing with the concern of overall delay. The link latency encodes the passive optical outages as well as the artificially proactive outages due to handover processes. The passive outages due to mobility can be predicted by LSTM, while the impact of handover outages is considered by Q-learning through the definition of rewards. Furthermore, the cumulative result of $r$ indicates the penalty due to the continuous outages during the handover process. This definition of reward unveils the motivation of the handovers from a seemingly stabler RF BS to an optical BS at higher risk but accompanied by potentially huge returns; and it makes the BS handle the balance between the outage risk and handover cost.

Data-driven methodologies rely on data statistics; therefore, once the mobility scenario changes, the spatio-temporal statistics change, and then the data-driven method needs to be refined to adapt to a new scenario. Thus, we investigate the performance of the data-driven handover framework in another scenario setting as the meeting room shown in Fig. 4.1b. We set the optical and RF transmission power for all links as 0.9 W and 0.01 W, respectively, and $R_T$ as 30 Mbit/s for all links. Using the same LSTM and Q-learning parameter set in the previous scenario, we retrain the BS controller, which leads to the results shown in Fig. 4.10. After only $10^4$ iterations, our data-driven approach outperforms the conventional handover methods remarkably,

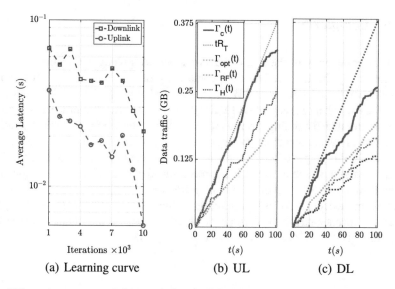

**Fig. 4.10** **a** Average network latency during the Q-learning process in the meeting room. After $10^4$ iterations, the latency become lower than in the conventional handover decisions. **b** and **c** Data traffic curves using the handover policy gained at the last iteration shown in **a**. In **b**, $\Gamma_{opt}$ and $\Gamma_H$ coincide [2]

and the average network latency from the introduced Q-learning algorithm can be reduced to 6 ms in uplink and 11 ms in downlink. The overall network latency of downlink has been improved by 98.4% compared to the instantaneous handover method in downlink HetNet, by 65.1% compared to the instantaneous handover in optical standalone downlink, and the improvement in uplink is 25.3% compared to the instantaneous handover in both HetNet and optical standalone network. For example, in such a meeting room for the result shown in Fig. 4.10b, c, we notice that the optical uplink holds less outages than the downlink, which causes $\Gamma_{opt}$ and $\Gamma_H$ in uplink to coincide since there always exist at least one light BS not suffering from outages. Therefore, the data-driven handover framework in the uplink learns specifically the optimal handover policy on choosing the most robust among the light BSs because no RF BS was needed. As such, the improvement in such uplink is, however, less than the downlink since the latency margin to be reduced in this uplink is less than that in the downlink.

This adaptation process takes about 32 seconds on a portable personal computer with Intel Core i7 processor for the Q-learning and 39 mins on a cluster with GeForce RTX 2080 Ti GPU for the deep-LSTM model training, which is a very short time when conducted on the BS server, so it can be arranged whenever the scenario gets modified due to a change in the room layout. Note that if we consider a longer training period until Q-learning converges, the performance can be further improved; this continued training process can be also carried out online after the initial deployment of the Q-learning in practice.

### 4.6.6   Complexity

The time complexity of the link evaluation stage with outage prediction is mostly contributed by the LSTM. The time complexity of LSTM is $O(W)$ [16], where $W$ is the number of weights and $W = \mathbb{P} \times (4 \times C \times (C + V + 1)) + C$, where $\mathbb{P}$ denotes the number of layers of LSTM. At the BS server side, this process duration is much shorter than the timing errors as well as $k$ when conducting the link evaluation. While providing an acceptable prediction accuracy, $W$ must be limited to reduce the time cost in this stage; otherwise, the actual equivalent timing delay error will be greatly enlarged, which is why it is not recommended to involve more complicated methods such as CNN [22] in the outage prediction procedure.

The complexity order of a Q-learning algorithm is $O(|\mathcal{S}|^2|\mathcal{A}|/(\epsilon_Q^3(1 - \gamma)^3))$ [23], where $| \cdot |$ corresponds to the cardinality of a space, and in this chapter, it approximates to $O(((V + 1)^2 M)^2 \times (V + 1)/(\epsilon_Q^3(1 - \gamma)^3))$.

## 4.7   Summary

In order to implement an optimal vertical handover decision policy in an RF-optical HetNet, this chapter covered the following topics:

- A data-driven handover framework that learns from the spatio-temporal dependency and decides the optimal handovers and link assignments was proposed. The core of the introduced framework is based on reinforcement learning (Q-learning) that takes advantage of the features of a Markov decision process (MDP) when learning from the mobile optical wireless channel, which results in an optimal vertical handover policy that minimizes the overall network latency and guarantees a target throughput for the UEs.
- The performance of the reinforcement learning-based vertical handover framework is further enhanced with the help of deep LSTM-based recurrent neural network (RNN) that learns the dependency in outage events from the channel status, yielding further insights on the channel estimation and more judicious state-action relevance.
- Extensive simulation results were performed using the generated mobile traces from Chap. 3 in a realistic indoor setting. Our results demonstrate the effectiveness of the introduced data-driven handover framework, which exhibits considerable improvement on the overall network latency and handover rate under indoor mobility for both uplink via RF and IR as well as downlink via RF and VLC.

# References

1. A. Al-Fuqaha et al., Internet of things: a survey on enabling technologies, protocols, and applications. IEEE Commun. Surv. Tutor. **17**(4), 2347–2376 (2015)
2. Z. Wu et al., Data-driven link assignment with QoS guarantee in mobile RF-optical HetNet-of-Things. IEEE Internet Things J. 1–1 (2020)
3. R. Zhang et al., Visible light communications in heterogeneous networks: paving the way for user-centric design. IEEE Wirel. Commun. **22**(2), 8–16 (2015). ISSN: 1536-1284
4. H. Haas et al., What is LiFi? J. Lightw. Technol. **34**, 1533–1544 (2016)
5. X. Wu, M. Safari, H. Haas, Access point selection for hybrid Li-Fi and Wi-Fi networks. IEEE Trans. Commun. **65**(12) (2017), 5375–5385. ISSN: 1558-0857
6. M.D. Soltani et al., Handover modeling for indoor Li-Fi cellular networks: the effects of receiver mobility and rotation, in *2017 IEEE Wireless Communications and Networking Conference (WCNC)* (2017), pp. 1–6
7. D. Wu, R. Negi, Effective capacity: a wireless link model for support of quality of service. IEEE Trans. Wirel. Commun. **24**(5), 630–643 (2003). ISSN: 1536-1276. http://ieeexplore.ieee.org/document/1210731/
8. F. Jin, R. Zhang, L. Hanzo, Resource allocation under delay-guarantee constraints for heterogeneous visible-light and RF femtocell. IEEE Trans. Wirel. Commun. **14**(2), 1020–1034 (2015). ISSN: 1536-1276. http://ieeexplore.ieee.org/document/6933944/
9. M. Hammouda et al., Link selection in hybrid RF/VLC systems under statistical queueing constraints. IEEE Trans. Wirel. Commun. **17**(4), 2738–2754 (2018). ISSN: 1536-1276. https://ieeexplore.ieee.org/document/8288604/
10. F. Wang et al., Efficient vertical handover scheme for heterogeneous VLC-RF systems. J. Opt. Commun. Netw. **7**(12), 1172 (2015). ISSN: 1943-0620. https://www.osapublishing.org/abstract.cfm?URI=jocn-7-12-1172
11. S. Zang et al., Managing vertical handovers in millimeter wave heterogeneous networks. IEEE Trans. Commun. **67**(2), 1629–1644 (2019). ISSN: 0090-6778. https://ieeexplore.ieee.org/document/8509624/
12. Z. Wang et al., Handover control in wireless systems via asynchronous multiuser deep reinforcement learning. IEEE Internet Things J. **5**(6), 4296–4307 (2018)
13. Y. Wang et al., Optimization of load balancing in hybrid LiFi/RF networks. IEEE Trans. Commun. **65**, 1708–1720 (2017)
14. D.A. Basnayaka, H. Haas, Design and analysis of a hybrid radio frequency and visible light communication system. IEEE Trans. Commun. **65**, 4334–4347 (2017)
15. Z. Wu et al., Efficient prediction of link outage in mobile optical wireless communications. IEEE Trans. Wirel. Commun. (2020) (under review)
16. S. Hochreiter, J. Schmidhuber, Long short-term memory. Neural Comput. **9**(8), 1735–1780 (1997)
17. D.P. Kingma, J. Ba, Adam: a method for stochastic optimization (2014), arXiv:1412.6980
18. X. Glorot, Y. Bengio, Understanding the difficulty of training deep feedforward neural networks, in Proceedings of the Thirteenth International Conference on Artificial Intelligence and Statistics (2010), pp. 249–256
19. A.M. Saxe, J.L. McClelland, S. Ganguli, Exact solutions to the nonlinear dynamics of learning in deep linear neural networks (2013), arXiv:1312.6120
20. C.J.C.H. Watkins, P. Dayan, Q-learning. Mach. Learn. **8**(3–4), 279–292 (1992)
21. M.D. Soltani et al., Modeling the random orientation of mobile devices: measurement, analysis and LiFi use case. IEEE Trans. Commun. **67**(3), 2157–2172 (2018)
22. K. He, J. Sun, Convolutional neural networks at constrained time cost, in *The IEEE Conference on Computer Vision and Pattern Recognition (CVPR)* (2015)
23. S.M. Kakade et al., On the sample complexity of reinforcement learning. PhD thesis. University College London, London, England (2003)

# Chapter 5
# Data-Driven Multi-homing Resource Allocation in Mobile 5G+ HetNets

**Abstract**  This chapter investigates resource management in 5G and beyond (5G+) heterogeneous networks (HetNets) for multi-homing connections. In such HetNets, channels from different networks vary at different timescales, which impacts resource allocation for multi-homing connections. We consider one application scenario for a 5G+ HetNet that integrates radio frequency (RF) and visible light base stations (BSs). In such a setting, RF channel gains vary faster than VLC channels due to small scale fading. Hence, to properly manage the HetNet resources at different timescales, we leverage multi-agent Q-learning to interact with the dynamics of wireless environments and present an online two-timescale power allocation policy that optimizes the transmit powers at the RF and optical BSs to ensure quality-of-service (QoS) satisfaction. Simulation results demonstrate the effectiveness of the introduced Q-learning based power allocation policy.

## 5.1  Introduction

As highlighted in Chap. 1, two kind of connections can be established in 5G and beyond (5G+) heterogeneous networks (HetNets), namely, multi-mode and multi-homing connections. In Chap. 4, we have investigated resource management for multi-mode connections in 5G+ HetNets via a data-driven handover framework. In this chapter, we investigate resource management for multi-homing connections. Specifically, with multi-homing connections, a user equipment (UE) aggregates the offered resources from different networks to support data rate hungry applications. The challenge herein setting is that: (1) Channels of different networks operating in different frequency bands vary at different timescales. (2) In multi-homing support, resource allocation in one network directly impacts the resource allocation from other networks. These two observations complicate the resource management for multi-homing connections in 5G+ HetNets. To investigate these challenges, we consider one application scenario of 5G+ HetNets that integrate radio frequency (RF) and visible light communication (VLC) base stations (BSs) for mobile indoor setting [1].

### 5.1.1 Background

For multi-homing connections in HetNets, decision making algorithms in different networks should operate at different timescales [2]. The authors in [3] examined a resource allocation problem for HetNets consisting of a wireless local area network (WLAN) and a cellular network where the two networks operate at different timescales. In [4], a two-timescale resource optimization framework for RF-optical HetNets with multi-mode users was introduced by employing the Lyapunov optimization method. However, the users' mobility and multi-homing capability were not taken into account in [4].

Resource allocation problems for RF-optical HetNets with mobile users exhibiting the multi-homing capability have been first studied in [1]. Unlike the case with multi-mode users in [4] where the achieved rate at a user is one of the rates from RF or VLC networks, the achieved rate at a multi-homing user is the sum of the rates from both the RF and VLC networks. Hence, it is important to carefully determine RF transmit powers in every time slot by considering the dynamics of the rates achieved by the VLC network while the power allocation at the VLC network is carried out at the beginning of each frame by taking the users' mobility into account.[1] Unfortunately, such a challenge is not well addressed in RF-optical 5G+ HetNets. In order to handle these issues, in [1] we adopted a data-driven approach by employing multi-agent Q-learning to develop an online two-timescale power allocation policy that identifies the transmit powers at the RF and optical BSs to satisfy the QoS requirements of mobile multi-homing users for indoor environments. We leverage the channel dataset created in Chap. 3 to implement the Q-learning-based power allocation policy.

### 5.1.2 Chapter Organization

In this chapter, we first describe the network model and introduce the problem formulation in Sect. 5.2. Then, we present the multi-agent Q-learning-based two-timescale resource allocation policy in Sect. 5.3. Finally, in Sect. 5.4 we present simulation results to investigate the performance of the proposed resource management policy.

## 5.2 Network Model and Problem Formulation

We investigate a wireless HetNet that consists of a single RF BS, $V$ optical BSs, and $U$ mobile users. It is assumed that each frame consists of $K$ time slots. In other words, the $n$-th frame $F_n$ contains $K$ time slots, i.e., $F_n = \{t_n, t_n + 1, \ldots, nK\}$ where $t_n = (n-1)K + 1$ means the time slot at the beginning of the $n$-th frame while $t$ stands for any time slot. Since the small-scale fading channels of RF links

---

[1] A single frame consists of multiple time slots.

vary faster than the channels of VLC links, optimizing the powers at the optical BSs in every time slot causes a computational overhead. Thus, we consider a power allocation framework for RF-optical HetNet operating at two different timescales. More specifically, the power allocations at the RF BS and optical BSs are conducted in every time slot and at the beginning of each frame, respectively.

We extend the mobility model in Chap. 3 to multi-user scenarios by including the avoidance force among users. The avoidance force between two users $\mathbf{F}_u(t)$ is given by

$$\mathbf{F}_u\left(\mathbf{d}_u(t), t\right) = \mathrm{m}\frac{(\mathbf{v}(t - \delta_\tau) - \mathbf{v}(t))\hat{\mathbf{n}}_u}{\delta_\tau\,|\mathbf{d}_u(t)|}, \tag{5.1}$$

in which $\mathbf{d}_u$ is the distance vector between the two users, and $\hat{\mathbf{n}}_u$ denotes an orthogonal unit vector against $\mathbf{d}_u$.

Mobile users present multi-homing capability that allows simultaneous association with both RF and optical BSs, but do not need vertical handovers among VLC BSs. The capacity of RF and optical channels follows the same definitions given in (4.2) and (4.3). Therefore, the total achievable rate at user $u$ becomes

$$R^{(u)}(t) = R_{\mathrm{RF}}^{(u)}(t) + R_{\mathrm{VLC}}^{(u,v(u,t))}(t), \tag{5.2}$$

where $v(u, t)$ denotes the optical BS supporting user $u$ in time slot $t$. Our goal is to develop an online algorithm that controls the transmit powers $\{P_{\mathrm{RF}}^{(u)}(t)\}$ and $\{P_{\mathrm{VLC}}^{(u,v(u,t))}(t_n)\}$ with the aim of providing a target rate $R_{\mathrm{T}}$ for each mobile user. Hence, the power allocation problem is formalized as follows:

$$\min_{\{P_{\mathrm{RF}}^{(u)}(t)\}, \{P_{\mathrm{VLC}}^{(u,v(u,t))}(t_n)\}} \sum_{t=1}^{\infty} \max_u \left| R^{(u)}(t) - R_{\mathrm{T}} \right| \tag{5.3}$$

$$\text{s.t.} \quad \sum_{u=1}^{U} P_{\mathrm{RF}}^{(u)}(t) \le P_{\mathrm{RF}}^{\max}$$

$$\sum_{u=1}^{U} P_{\mathrm{VLC}}^{(u,v(u,t))}(t_n) \le P_{\mathrm{VLC}}^{\max} \; \forall v,$$

where $P_{\mathrm{RF}}^{\max}$ and $P_{\mathrm{VLC}}^{\max}$ respectively account for the available powers at the RF and optical BSs. The formulation in (5.3) aims to specify the power allocations that satisfy the users' target rate by minimizing the absolute deviation between the achieved rate and the target rate for the worst user while accounting for the power limits at the RF and VLC BSs. The achieved rate $R^{(u)}(t)$ is described by (5.2).

## 5.3  Two-Timescale Power Allocation

In this section, we introduce an online two-timescale power allocation policy in which the power allocations at the RF and optical BSs are carried out in every time slot and at the beginning of each frame, respectively. In order to solve the optimization problem in (5.3) by taking the dynamics of the channels and users' mobility into account, each BS identifies its transmit power by employing a Q-learning-based approach. More specifically, each BS acts as an agent, and hence each BS observes its current state and chooses an action, i.e., transmit power, following its own decision policy that minimizes the expected discounted cumulative cost. Here, the state and action for the RF BS (or optical BSs) are determined in every time slot (or frame).

### 5.3.1  Definitions of State, Action and Cost

Define the state space of optical BS $v$ as

$$\mathcal{S}_{\text{VLC}}^{(v)} = \mathcal{S}_{\text{VLC}}^{(1,v)} \times \mathcal{S}_{\text{VLC}}^{(2,v)} \times \cdots \times \mathcal{S}_{\text{VLC}}^{(u,v)}, \tag{5.4}$$

where $\mathcal{S}_{\text{VLC}}^{(u,v)}$ indicates the state space of the link between optical BS $v$ and user $u$, and $\times$ is the Cartesian product. Note that when user $u$ is not located in the coverage region of optical BS $v$ at the beginning of the $n$-th frame, the achievable rate from optical BS $v$, $R_{\text{VLC}}^{(u,v)}(t)$, is always zero regardless of the allocated power for the user $P_{\text{VLC}}^{(u,v)}(t_n)$. Since the goal is to achieve the target rate $R_{\text{T}}$ as shown in (5.3), by considering the relationship between the achievable rate $R_{\text{VLC}}^{(u,v)}(t)$ and $R_{\text{T}}$, the representation of $\mathcal{S}_{\text{VLC}}^{(u,v)}$ is given by

$$\mathcal{S}_{\text{VLC}}^{(u,v)} = \begin{cases} s_{\text{VLC}}^{(0)}, & \text{if } R_{\text{VLC}}^{(u,v)}(t_n) = 0 \\ s_{\text{VLC}}^{(1)}, & \text{if } 0 < R_{\text{VLC}}^{(u,v)}(t_n) < R_{\text{T}} \\ s_{\text{VLC}}^{(2)}, & \text{otherwise.} \end{cases} \tag{5.5}$$

Also, we denote the state space of the RF BS as

$$\mathcal{S}_{\text{RF}} = \mathcal{S}_{\text{RF}}^{(1)} \times \mathcal{S}_{\text{RF}}^{(2)} \times \cdots \times \mathcal{S}_{\text{RF}}^{(U)}. \tag{5.6}$$

In each time slot $t$, to solve the problem in (5.3), the RF BS controls its transmit power with the aim of minimizing the difference between the target rate $R_{\text{T}}$ and the achieved rate by taking the rate achieved by the optical BSs into account. Thus, based on the comparison between the total achievable rate of user $u$ and $R_{\text{T}}$, $\mathcal{S}_{\text{RF}}^{(u)}$ is defined follows:

$$\mathcal{S}_{\text{RF}}^{(u)} = \begin{cases} s_{\text{RF}}^{(1)}, & \text{if } R_{\text{RF}}^{(u)}(t) + R_{\text{VLC}}^{(u,v(u,t_n))}(t_n) < R_{\text{T}} \\ s_{\text{RF}}^{(2)}, & \text{otherwise.} \end{cases} \tag{5.7}$$

Let us define sets of transmit powers at the optical BSs as

$$\mathcal{P}_{\text{VLC}} = \{P_{\text{VLC},1}, P_{\text{VLC},2}, \ldots P_{\text{VLC},V_P}\}, \tag{5.8}$$

and RF BS as

$$\mathcal{P}_{\text{RF}} = \{P_{\text{RF},1}, P_{\text{RF},2}, \ldots P_{\text{RF},R_P}\}, \tag{5.9}$$

where $V_P$ and $R_P$ indicate the numbers of power levels at the optical and RF BSs, respectively. Here, $P_{\text{VLC},i} \leq P_{\text{VLC}}^{\max}$ and $P_{\text{RF},i} \leq P_{\text{RF}}^{\max}$.

Then, we have the action space of optical BS $v$ expressed as

$$\mathcal{A}_{\text{VLC}}^{(v)} = \{\mathbf{a}_1^{(v)}, \mathbf{a}_2^{(v)}, \ldots, \mathbf{a}_{V_A}^{(v)}\}, \tag{5.10}$$

where $\mathbf{a}_i^{(v)} \triangleq [a_i^{(1,v)} \ a_i^{(2,v)} \ldots a_i^{(U,v)}]$ is a vector with size $U$. Here, $a_i^{(u,v)} \in \mathcal{P}_{\text{VLC}}$ means the allocated power for user $u$ and $V_A$ represents the number of combinations of $\{a_i^{(u,v)}\}$ that fulfill the power constraint $\sum_{u=1}^{U} a_i^{(u,v)} \leq P_{\text{VLC}}^{\max}$. In the same manner, the action space of the RF BS becomes

$$\mathcal{A}_{\text{RF}} = \{\mathbf{a}_1, \mathbf{a}_2, \ldots, \mathbf{a}_{R_A}\}, \tag{5.11}$$

where $\mathbf{a}_i \triangleq [a_i^{(1)} \ a_i^{(2)} \ldots a_i^{(U)}]$ is a vector with size $U$, $a_i^{(u)} \in \mathcal{P}_{\text{RF}}$ stands for the allocated power for user $u$, and $R_A$ designates the number of combinations of $\{a_i^{(u)}\}$ that satisfy the condition $\sum_{u=1}^{U} a_i^{(u)} \leq P_{\text{RF}}^{\max}$.

Since the goal of the power allocation policy is to provide the target rate $R_T$ for each user as expressed in (5.3), we compute the cost based on the differences between $R_T$ and the achieved rates from the RF and optical BSs. For optical BS $v$ and the $n$-th frame, we define $E_{\text{VLC}}^{(v)}(t_n)$ and $D_{\text{VLC}}^{(v)}(t_n)$ as

$$E_{\text{VLC}}^{(v)}(t_n) = \left(\max_u\{R_{\text{VLC}}^{(u,v)}(t_n)\} - R_T\right)_+^{\theta_{\text{VLC}}^E}, \tag{5.12}$$

$$D_{\text{VLC}}^{(v)}(t_n) = \left(R_T - \min_u\{R_{\text{VLC}}^{(u,v)}(t_n)\}\right)_+^{\theta_{\text{VLC}}^D}, \tag{5.13}$$

where $\theta_{\text{VLC}}^E$ and $\theta_{\text{VLC}}^D$ are constants, and $(x)_+ \triangleq \max(x, 0)$. Here, $E_{\text{VLC}}^{(v)}(t_n)$ and $D_{\text{VLC}}^{(v)}(t_n)$ respectively indicate the *excess* and *deficiency* of the achieved rate from optical BS $v$ over the target rate $R_T$. Then, when optical BS $v$ takes an action in time slot $t_n$, the cost is given by

$$C_{\text{VLC}}^{(v)}(t_n) = E_{\text{VLC}}^{(v)}(t_n) + D_{\text{VLC}}^{(v)}(t_n). \tag{5.14}$$

In each time slot, the RF BS identifies its transmit power by considering the rates achieved by the optical BSs $\{R_{\text{VLC}}^{(u,v(u,t))}(t)\}$. Note that $R_{\text{VLC}}^{(u,v(u,t))}(t)$ can significantly vary with $t$ depending on whether user $u$ is located in the coverage region of the optical BSs in time slot $t$. To reduce the impact of the dynamics of $R_{\text{VLC}}^{(u,v(u,t))}(t)$, we consider the average achieved rate from the optical BSs into account and define the excess and deficiency for the RF BS in time slot $t$ as follows:

$$E_{\text{RF}}(t) = \left( \max_{u} \{ R_{\text{RF}}^{(u)}(t) + \Gamma^{(u)}(t) \} - R_{\text{T}} \right)_{+}^{\theta_{\text{RF}}^{E}}, \tag{5.15}$$

$$D_{\text{RF}}(t) = \left( R_{\text{T}} - \min_{u} \{ R_{\text{RF}}^{(u)}(t) + \Gamma^{(u)}(t) \} \right)_{+}^{\theta_{\text{RF}}^{D}}, \tag{5.16}$$

where $\theta_{\text{RF}}^{E}$ and $\theta_{\text{RF}}^{D}$ are constants, and

$$\Gamma^{(u)}(t) = \frac{1}{t} \sum_{\tau=1}^{t} \left( R_{\text{VLC}}^{(u,v(u,\tau))}(\tau) \right) \tag{5.17}$$

denotes the average total achieved rate from the optical BSs. The cost function of the RF BS is expressed as:

$$C_{\text{RF}}(t) = E_{\text{RF}}(t) + D_{\text{RF}}(t). \tag{5.18}$$

It is interesting to observe that the terms $\theta_{\text{VLC}}^{E}, \theta_{\text{RF}}^{E}, \theta_{\text{VLC}}^{D}$ and $\theta_{\text{RF}}^{D}$ can be interpreted as the factors that determine the tolerance on the excess and deficiency. Since $E_{\text{VLC}}^{(v)}(t_n), D_{\text{VLC}}^{(v)}(t_n), E_{\text{RF}}(t)$ and $D_{\text{RF}}(t)$ represent the differences between the achieved rates and the target rate, the users can achieve the target rate when the cost functions in (5.14) and (5.18) are minimized.

### 5.3.2  Multi-agent Q-learning-based Power Allocation

In Q-learning, by executing an action $a$ at a state $s$, an agent learns the expected discounted cumulative cost $Q(s, a)$, which is called Q-value. For a given current state $s$, the agent chooses its action $a$ that exhibits the lowest Q-value among all possible actions in its action space $\mathcal{A}$:

$$a = \arg\min_{\tilde{a} \in \mathcal{A}} Q(s, \tilde{a}). \tag{5.19}$$

When the state in the beginning of the $n$-th frame is $s^{(v)}$, optical BS $v$ receives the cost in (5.14) after selecting the action $\mathbf{a}^{(v)}$, and then optical BS $v$ updates its Q-value $Q_{\text{VLC}}^{(v)}(s^{(v)}, \mathbf{a}^{(v)})$ for this state-action pair as follows:

$$Q_{\text{VLC}}^{(v)}(s^{(v)}, \mathbf{a}^{(v)}) \leftarrow (1 - \alpha_V) Q_{\text{VLC}}^{(v)}(s^{(v)}, \mathbf{a}^{(v)})$$
$$\alpha_V \left( C_{\text{VLC}}^{(v)}(t_n) + \gamma_V \min_{\mathbf{a}^{(v)} \in \mathcal{A}_{\text{VLC}}^{(v)}} Q_{\text{VLC}}^{(v)}(\hat{s}^{(v)}, \mathbf{a}^{(v)}) \right), \tag{5.20}$$

where $\alpha_V$ and $\gamma_V$ respectively stand for the learning rate and discount factor for the optical BSs, and $\hat{s}^{(v)}$ indicates the next state. In time slot $t$, the Q-value at the RF BS $Q_{RF}(s, \mathbf{a})$ is updated as follows:

$$Q_{RF}(s, \mathbf{a}) \leftarrow (1 - \alpha_R)Q_{RF}(s, \mathbf{a}) + \alpha_R\left(C_{RF}(t) + \gamma_R \min_{\mathbf{a} \in \mathcal{A}_{RF}} Q_{RF}(\hat{s}, \mathbf{a})\right), \quad (5.21)$$

where $\alpha_R$ and $\gamma_R$ account for the learning rate and discount factor for the RF BS, respectively, and $\hat{s}$ means the next state. The details of the online two-timescale power allocation policy are provided in Algorithm 7.

---

**Algorithm 7. Two-Timescale Power Allocation Policy**

---

1. Set $t = 1$ and $n = 1$. For all state-action pairs, initialize the Q-values $Q_{RF}(s, \mathbf{a}) = 0$ and $Q_{VLC}^{(v)}(s^{(v)}, \mathbf{a}^{(v)}) = 0$.
2. **loop**
3.   **if** $t = t_n$ **then**
4.     **for** $v := 1$ to $V$ **do**
5.       Generate a random number $r$ from $\mathcal{M}$.
6.       **if** $r \leq \epsilon_V$ **then**
7.         Select an action $\mathbf{a}^{(v)}$ randomly from $\mathcal{A}_{VLC}^{(v)}$.
8.       **else**
9.         Select $\mathbf{a}^{(v)}$ according to $\arg\min_{\mathbf{a}^{(v)} \in \tilde{\mathcal{A}}_{VLC}^{(v)}} Q_{VLC}^{(v)}(s^{(v)}, \mathbf{a}^{(v)})$.
10.     **end if**
11.   **end for**
12.     Execute $\{\mathbf{a}^{(v)}\}$ and receive the costs $\{C_{VLC}^{(v)}(t_n)\}$ in (5.14).
13.     Observe the next states $\{\hat{s}^{(v)}\}$.
14.     Update $Q_{VLC}^{(v)}(s^{(v)}, \mathbf{a}^{(v)})$ according to (5.20).
15.     Set $s^{(v)} = \hat{s}^{(v)}$ for $\forall v$ and $n = n + 1$.
16.   **end if**
17.   Generate a random number $r$ from $\mathcal{M}$.
18.   **if** $r \leq \epsilon_R$ **then**
19.     Select an action $\mathbf{a}$ randomly from $\mathcal{A}_{RF}$.
20.   **else**
21.     Select $\mathbf{a}$ according to $\arg\min_{\mathbf{a} \in \mathcal{A}_{RF}} Q_{RF}(s, \mathbf{a})$
22.   **end if**
23.   Execute $\mathbf{a}$ and receive the cost $C_{RF}(t)$ in (5.18)
24.   Observe the next state $\hat{s}$.
25.   Update $Q_{RF}(s, \mathbf{a})$ according to (5.21).
26.   Set $s = \hat{s}$ and $t = t + 1$.
27. **end loop**

---

In Algorithm 7, the optical BSs determine their transmit powers at the beginning of each frame $t_n$ and the RF BS updates its transmit power in every time slot $t$. More specifically, each optical BS independently chooses its action based on its state and

Q-value in (5.5) and (5.20). Then, by leveraging the information of the achieved rates from the optical BSs $\{R_{\text{VLC}}^{(u,v(u,t))}(t)\}$, the RF BS selects its action from its state and Q-value in (5.7) and (5.21). Here, the $\epsilon$-greedy strategy [5] is adopted where $\mathcal{M}$ is a uniform distribution over the interval $[0, 1]$. Therefore, the RF BS (or optical BS $v$) chooses a random action with probability $\epsilon_R$ (or $\epsilon_V$) and selects an action based on its Q-values with probability $1 - \epsilon_R$ (or $1 - \epsilon_V$) where $\epsilon_R \in [0, 1]$ and $\epsilon_V \in [0, 1]$. Note that if $S_{\text{VLC}}^{(u,v)} = 0$, user $u$ is not located in the coverage region of optical BS $v$, and hence allocating power to user $u$ has no impact on the rate. In this regard, when $S_{\text{VLC}}^{(u,v)} = 0$ in time slot $t_n$, optical BS $v$ selects its action from the set $\tilde{\mathcal{A}}_{\text{VLC}}^{(v)}$ where $\tilde{\mathcal{A}}_{\text{VLC}}^{(v)}$ consist of the vectors $\{\mathbf{a}_i^{(v)}\}$ with $a_i^{(u,v)} = 0$. In Algorithm 7, the VLC and RF BSs determine their powers based on Q-values in (5.20) and (5.21), which are based on the cost functions in (5.14) and (5.18). Thus, from Algorithm 7, the users can attain the target rate $R_T$ that coincides with the object of the problem in (5.3).

### 5.3.3  Complexity

The complexity order of a Q-learning algorithm is $O(|\mathcal{S}|^2|\mathcal{A}|/(\epsilon^3(1 - \gamma)^3))$ where $\mathcal{S}, \mathcal{A}, \epsilon, \gamma$ and $|\cdot|$ respectively mean the state space, action space, $\epsilon$-greedy parameter, discount factor, and the cardinality of a space [6]. In Algorithm 7, the VLC and RF BSs update their Q-values in series, and hence, the complexity order of the power allocation policy is given by $O\left( \max\left( \max_{v=1,...,V} \frac{|S_{\text{VLC}}^{(v)}|^2|\mathcal{A}_{\text{VLC}}^{(v)}|}{\epsilon_V^3(1-\gamma_V)^3}, \frac{|S_{\text{RF}}|^2|\mathcal{A}_{\text{RF}}|}{\epsilon_R^3(1-\gamma_R)^3}\right)\right)$ where the sizes of the state space and action space are given by $|S_{\text{VLC}}^{(v)}| = 3^U$, $|S_{\text{RF}}| = 2^U$, $|\mathcal{A}_{\text{VLC}}^{(v)}| = V_A$, and $|\mathcal{A}_{\text{RF}}| = R_A$.

## 5.4  Numerical Results

We illustrate numerical simulation results to validate Algorithm 7. We adopt the office environment setup in Fig. 4.1a of Chap. 4. The bandwidth of both VLC and RF BSs is set as 20 MHz. The maximum signal power is $P_{\text{VLC}}^{\max} = 3$ W for optical BS, and $P_{\text{RF}}^{\max} = 10$ dBm for the RF BS. Unless otherwise stated, we set $U = 2$, $\gamma_R = \gamma_V = 0.9$, $\alpha_R = \alpha_V = 0.5$, $\epsilon_R = 0.1$, $\epsilon_V = 0.2$. The durations of each time slot and frame are respectively $d_T = 10$ ms and $d_F = 200$ ms, and

$$P_{\text{RF},i} = \frac{(i-1)}{R_P - 1}P_{\text{RF}}^{\max}, \tag{5.22}$$

$$P_{\text{VLC},i} = \sqrt{\frac{(i-1)}{V_P - 1}(P_{\text{VLC}}^{\max})^2}, \tag{5.23}$$

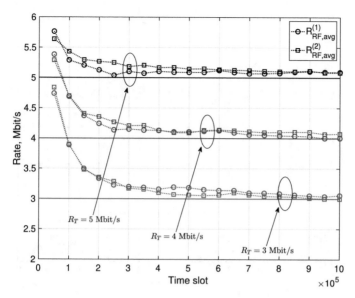

**Fig. 5.1** Average user data rate of RF standalone network, where $\theta_{RF}^E = 1$ and $\theta_{RF}^D = 2$

where $R_P = V_P = 31$. The average rates at time slot $T$ from the RF and VLC networks correspond to

$$R_{RF,avg}^{(u)}(T) = \frac{1}{T}\sum_{t=1}^{T} R_{RF}^{(u)}(t),$$  (5.24)

$$R_{VLC,avg}^{(u)}(T) = \frac{1}{T}\sum_{t=1}^{T} R_{VLC}^{(u,v(u,t))}(t).$$  (5.25)

We first examine the performance of the standalone networks. In the RF standalone mode, $R_{VLC}^{(u,v)}(t) = 0$ for the state definition as well as the cost. If we let $\theta_{RF}^E = \theta_{RF}^D$, due to the small-scale fading in RF channel, the QoS cannot be satisfied since the penalty of deficiency is not enough. As such, we should tune the parameter $\theta^D$ for enhancing the penalty for RF to tackle this problem so that we can get the results shown in Fig. 5.1.

Figure 5.2 presents the performance of the Q-learning algorithm for VLC BSs. However, for VLC, if the achieved rate cannot satisfy the target rate, there will always be some of the BSs out of view.

Then, we examine the performance of RF-VLC HetNets. We set 3 target rates to examine the presented algorithm. When a low data rate $R_T = 40$ Mbit/s is set, we have $\theta_{VLC}^E = 3, \theta_{VLC}^D = 1, \theta_{RF}^E = 1$, and $\theta_{RF}^D = 2$ to suppress the use of optical BSs. When a medium data rate $R_T = 50$ Mbit/s is set, we have $\theta_{VLC}^E = 1, \theta_{VLC}^D = 1, \theta_{RF}^E = 1$, and $\theta_{RF}^D = 2$ to encourage the use of RF BS to supplement the achieved rate towards the

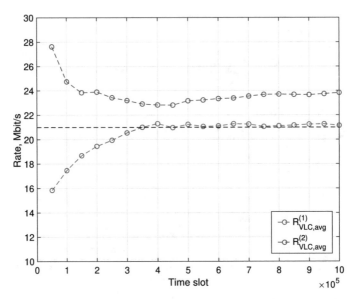

**Fig. 5.2**  Average user data rate of VLC standalone network, where $\theta_{\text{VLC}}^{E} = 1$ and $\theta_{\text{VLC}}^{D} = 1$

target rate. When a high data rate $R_{\text{T}} = 60$ Mbit/s is set, we have $\theta_{\text{VLC}}^{E} = 1, \theta_{\text{VLC}}^{D} = 2$, $\theta_{\text{RF}}^{E} = 1$, and $\theta_{\text{RF}}^{D} = 2$ to stimulate both networks to cooperatively achieve the target rate. In Fig. 5.3, we evaluate the average rates achieved by the power allocation policy as a function of the number of time slots $T$ with different values of the target rate $R_{\text{T}}$. Here, the average rate of user $u$ is

$$R_{\text{avg}}^{(u)} = R_{\text{RF,avg}}^{(u)} + R_{\text{VLC,avg}}^{(u)}. \tag{5.26}$$

From Fig. 5.3, we can observe that the presented algorithm successfully satisfies the QoS requirements of the users. This indicates that the RF network well compensates the required additional rates that cannot be provided solely by the VLC network due to the fact that the power allocation at the VLC network is conducted only at the beginning of each frame, i.e., $t_n$ for all $n$.

In this chapter, from the perspective of entire system, we are performing a distributed multi-agent Q-learning. As in Algorithm 7, the RF BS sets its allocated power after all VLC BSs have completed their power allocation, which implies a fluctuation in the target rate of the RF BS. Hence, the data rate achieved by the RF BS ranges in a very narrow field, since it cannot link its action with its state independently, which is coupled with the rate achieved through the VLC BS in advance. On the other hand, the VLC BS allocates its power towards achieving the target rate from the beginning, so it is very likely to attain a relatively high data rate. Consequently, with the consideration of the coupling with the RF BS, the VLC BS component

**Fig. 5.3** Average rates of
users as a function of the
number of time slots $T$

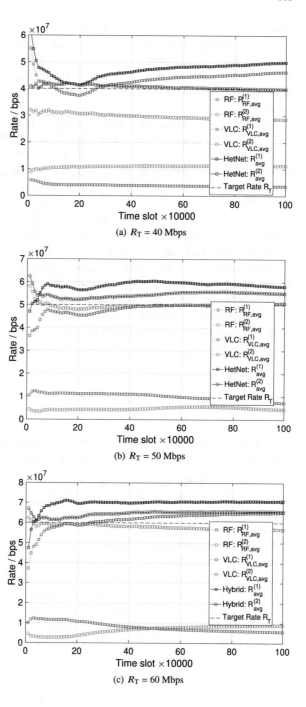

(a) $R_T = 40$ Mbps

(b) $R_T = 50$ Mbps

(c) $R_T = 60$ Mbps

results in a total rate that effortlessly suffices the target. Such an allocation strategy is robust, where each BS chooses actions based on its own state without considering the fluctuations of data rates through other BSs. However, as this is a two-timescale strategy and the BSs are making decisions one by one, the Q-learning at the RF BS reduces to a centralized Q-leaning since it observes all other BSs and makes its final decision.

Meanwhile, we would like to add a remark for better understanding the decision on $\theta$. This relates to the subject of optimization for a multi-agent Q-learning as well. First of all, the selections of $\theta^E$ and $\theta^D$ are not entirely isolated, since the ratio between $\theta^E$ and $\theta^D$ affects the trend of data rate with respect to iteration. In other words, rather than choosing a set of specific values, it is better to choose a proper ratio. Of course, one can tune the set of $\theta$ according to a specific $R_T$ or within a certain range of $R_T$ by an optimization process. However, we have not achieved a comprehensive and efficient approach for such tuning yet, since dealing with the relationship among $\theta$, $R_T$, and the actual data rate is analytically infeasible.

Multi-agent Q-learning involves Markov gaming among the agents and is still facing many challenges, especially in a highly dynamic environment. The future wireless networks are bound to meet the challenges of joint operation and optimization of multiple BSs operating at various frequency bands. Therefore, the method we resort to in this chapter, which supports a multi-timescale and multi-homing HetNet via multi-agent Q-learning, can be seen as a starting point for further studies.

## 5.5  Summary

This chapter investigated the power allocation problem for RF-optical HetNets with multi-homing users. The major findings of this chapter are summarized as follows:

- An online two-timescale power allocation algorithm, which utilizes multi-agent Q-learning to learn the dynamics of the mobile channels, was presented. The proposed power allocation strategy successfully supports the users while satisfying in the same time their QoS requirements.
- A multi-agent cooperation scheme was proposed to link the status of each agent at BS side properly; thus, every agent makes decision based on the balance between its own service quality and the observation on other BSs.
- A tolerance factor was proposed to determine the allowable variation on the excess and deficiency of QoS to ensure that the agents can learn to achieve the goal efficiently. Such a tolerance factor also helps the Q-learning tackle the small-scale fading problem in RF channel and optimize the cooperation among the BSs.

# References

1. J. Kong et al., Q-learning based two-timescale power allocation for multi-homing hybrid RF/VLC networks. IEEE Wirel. Commun. Lett. **9**(4), 443–447 (2020)
2. H.S. Chang et al., Multitime scale Markov decision processes. IEEE Trans. Autom. Control **48**, 976–987 (2003)
3. A. Tharaperiya Gamage, H. Liang, X. Shen, Two time-scale cross-layer scheduling for cellular/WLAN interworking. IEEE Trans. Commun. **62**, 2773–2789 (2014)
4. W. Wu, F. Zhou, Q. Yang, Adaptive network resource optimization for heterogeneous VLC/RF wireless networks. IEEE Trans. Commun. **66**, 5568–5581 (2018)
5. R.S. Sutton, A.G. Barto, *Reinforcement Learning: An Introduction* (MIT Press, Boca Raton, 1998)
6. S.M. Kakade et al., On the sample complexity of reinforcement learning. PhD thesis. University College London, London, England (2003)

# Chapter 6
# Conclusions

**Abstract** This brief has covered both planning and operational aspects of 5G and beyond (5G+) heterogeneous networks (HetNets). While the application scenario presented in this brief consider integration of radio frequency (RF) and optical (visible light communication (VLC) and infrared light (IR) communication) networks, many of the concepts introduced here can be applied to integrate other high frequency bands such as millimeter wave (mmWave) and Terahertz bands. Below, we summarize the main findings in this brief and we highlight future research directions.

## 6.1  Summary

This section summarizes the main contributions and conclusions presented in this brief.

### 6.1.1  Holistic Joint Planning Strategy for 5G+ HetNets

In this brief, we first focused on energy efficient deployment of standalone VLC networks and RF-optical HetNets that satisfy a service requirement on outage probability and present low power consumption. Firstly, by using stochastic geometry results, tight approximations of the outage probability of VLC networks were presented. The outage probability expressions presented herein brief are applicable to an arbitrary field-of-view (FOV) at the receiver's photo-diode (PD) and present low computational complexities. Then, by leveraging such approximations, an algorithm was developed to identify optimal base station (BS) density in standalone VLC to minimize the area power consumption (APC) while satisfying an outage probability constraint. In addition, new algorithms to develop energy efficient RF-optical Het-Nets were presented to optimize the densities of RF macro and small BSs and VLC BSs. The presented algorithms assume low (computational) complexity burden and high accuracy. Numerical simulations corroborated the efficiency of the algorithms and validated the fact that RF-optical HetNets can achieve more strict requirements

Z.-Y. Wu et al., *Efficient Integration of 5G and Beyond Heterogeneous Networks*, https://doi.org/10.1007/978-981-15-6938-8_6

on outage probability using lower power consumption compared to standalone RF networks and standalone VLC networks.

### 6.1.2 Mobile Channel Dataset for AI-Empowered Network Operation

The high dynamics and complex settings expected in 5G+ HetNets motivate adopting data-driven approaches for efficient management of network resources. However, such artificial intelligence (AI)-empowered strategies require access to rich datasets of wireless mobile channels, which currently are not available. Towards this objective, we introduced a framework to generate indoor human mobility so as to tackle the shortage and difficulty in collecting indoor mobility datasets. The presented mobility model captures human aspects by integrating macro and micro mobility patterns. The macro pattern is dominated by a semi-Markov renewal process subject to bounded Lévy-walk and return regularity, and the micro pattern is featured by the behaviors of path selection, steering, and orientation of user equipment (UE). This mobility model was matched with real-world measurements. The presented mobility model was then used to generate a dataset of RF and optical channels and can be used further to generate channel datasets for other high frequency bands such as millimeter wave (mmWave) and Terahertz.

Based on the generated mobile channel dataset, we have presented deep insights on indoor mobile optical channels. As for the line-of-sight (LOS) links, the overall channel gain follows multiple-peak Nakagami distributions. As for the non-LOS (NLOS) links, the channel gain distributions in the uplink are subject to Gamma and Nakagami distributions, but downlinks follow Burr distributions. The uplink bandwidth in NLOS follows the generalized log-logistic distribution, while the downlink NLOS bandwidth follows multiple-peak log-logistic distributions. All these distributions are found to be space-time-dependent. Furthermore, the LOS outage probability exhibits an apparent spatial symmetry. From the perspective of time, the probability distributions of outage and channel gain are basically stationary while user is wandering indoors, and are dominated by the trajectories. However, along the way towards the first entrance or exit point driven by the macro mobility patterns, these distributions suffer from a strong disturbance that evolves over time. From a spatial perspective, the distributions of the channel gain are still dominated by spatial structures.

### 6.1.3 Data-Driven Operation for 5G+ HetNets

In this brief, we have considered two connection types in 5G+ HetNets, namely multi-mode and multi-homing connections. For multi-mode connections, our analysis on the rate of conventional (channel gain-based) handover technique under environment-confined mobility yielded ill-conditioned results owing to frequent outages in optical links, which motivates further investigations on more intelligent resource management (handover) strategies. The optimization objective for quality-of-service (QoS)-guaranteed link assignment aims to minimize the overall handover frequency and network latency. Using the created mobile channel datasets, we introduced an AI-enabled model of progressive changes in mobile optical channel status, which can predict burst outages in LOS optical links using deep long-short-term-memory (LSTM) recurrent neural networks (RNNs). Through proper evaluation of optical link status with the help of outage predictions, a Q-learning-based handover strategy was presented and shown to exhibit a considerable reduction in the overall latency and handover rate. Such an intelligent handover strategy balances data traffic, outage risk, and cost of handover delay by simply observing the channel gain instead of digging into the user's movement details.

For multi-homing connections, we investigated the power allocation problem. Since RF channels change more rapidly than VLC channels, we introduced an AI-based algorithm for online two-timescale power allocation. Such an algorithm benefits from Q-learning and aims to learn the dynamics of the RF and optical wireless channels to carry out power allocations that satisfy QoS requirements. The presented two-timescale policy makes use of the created mobile channel dataset presented in this brief.

## 6.2 Future Research Directions

In this brief, we mainly focused on investigating the possibility and benefits of integrating efficiently new electromagnetic media into 5G+ HetNets. Our objective in this brief was to convey our understanding of the possible ways to solve the challenges expected in 5G+ HetNets. Our proposed methodology along with the created datasets will help to open up an AI-enabled perspective to solve more complex problems in 5G+ HetNets.

While our proposed methodology has been applied for integration of VLC and IR BSs with RF BSs, such a methodology is still applicable to other high frequency bands including mmWave and Terahertz. However, these new media will also present new challenges that need further considerations. This is mainly due to the fact that the BSs operating at different wavelengths present different coverage and penetration abilities. One obvious challenge that requires further investigation is the strong interference present among adjacent BSs, which can no longer be ignored. This may require a higher level of intelligent cooperation among networks that optimizes band assignment or user scheduling to maintain acceptable QoS.

In addition, 5G+ HetNets may include not only two types of media, which calls for cooperation among multiple bands, for instance, RF, mmWave, visible light, and Terahertz. With the presence of more candidate links, more aggressive handovers for multi-mode users and more unbalanced recourse allocation for multi-homing

# Index

Printed in the United States
by Baker & Taylor Publisher Services